Lecture Notes in Computer Science

Lecture Notes in Artificial Intelligence 15462
Founding Editor

Jörg Siekmann

Series Editors

Randy Goebel, *University of Alberta, Edmonton, Canada*
Wolfgang Wahlster, *DFKI, Berlin, Germany*
Zhi-Hua Zhou, *Nanjing University, Nanjing, China*

The series Lecture Notes in Artificial Intelligence (LNAI) was established in 1988 as a topical subseries of LNCS devoted to artificial intelligence.

The series publishes state-of-the-art research results at a high level. As with the LNCS mother series, the mission of the series is to serve the international R & D community by providing an invaluable service, mainly focused on the publication of conference and workshop proceedings and postproceedings.

Petia Koprinkova-Hristova · Nikola Kasabov
Editors

Artificial Intelligence: Methodology, Systems, and Applications

19th International Conference, AIMSA 2024
Varna, Bulgaria, September 18–20, 2024
Proceedings

Editors
Petia Koprinkova-Hristova
Institute of Information and Communication
Technologies (IICT), Bulgarian Academy
of Sciences
Sofia, Bulgaria

Nikola Kasabov
Auckland University of Technology
Auckland, New Zealand

Institute of Information and Communication
Technologies (IICT), Bulgarian Academy
of Sciences
Sofia, Bulgaria

ISSN 0302-9743 ISSN 1611-3349 (electronic)
Lecture Notes in Artificial Intelligence
ISBN 978-3-031-81541-6 ISBN 978-3-031-81542-3 (eBook)
https://doi.org/10.1007/978-3-031-81542-3

LNCS Sublibrary: SL7 – Artificial Intelligence

© The Editor(s) (if applicable) and The Author(s), under exclusive license
to Springer Nature Switzerland AG 2025

This work is subject to copyright. All rights are solely and exclusively licensed by the Publisher, whether the whole or part of the material is concerned, specifically the rights of translation, reprinting, reuse of illustrations, recitation, broadcasting, reproduction on microfilms or in any other physical way, and transmission or information storage and retrieval, electronic adaptation, computer software, or by similar or dissimilar methodology now known or hereafter developed.
The use of general descriptive names, registered names, trademarks, service marks, etc. in this publication does not imply, even in the absence of a specific statement, that such names are exempt from the relevant protective laws and regulations and therefore free for general use.
The publisher, the authors and the editors are safe to assume that the advice and information in this book are believed to be true and accurate at the date of publication. Neither the publisher nor the authors or the editors give a warranty, expressed or implied, with respect to the material contained herein or for any errors or omissions that may have been made. The publisher remains neutral with regard to jurisdictional claims in published maps and institutional affiliations.

This Springer imprint is published by the registered company Springer Nature Switzerland AG
The registered company address is: Gewerbestrasse 11, 6330 Cham, Switzerland

If disposing of this product, please recycle the paper.

Preface

Since its first edition in 1984 the International Conference on Artificial Intelligence: Methodology, Systems, Applications (AIMSA) has been held biennially in Bulgaria. The conference always covers a wide range of AI topics aiming at a wide audience of both AI practitioners and theorists. The conference has always responded to up-to-date issues and contemporary topics in AI.

After the last edition in 2018, the conferences in 2020 and 2022 were canceled due to the COVID-19 pandemic. This year AIMSA was revived in a hybrid mode, both in person and on-line, and the venue was the beautiful resort of Varna on the Black See.

The 19th edition focused on the following topics:

- Innovative methods for efficient and sustainable AI
- Systems for the development and implementation of efficient and sustainable AI
- Novel efficient and sustainable AI applications

This year 18 papers out of 23 submissions were accepted as full papers after rigorous single-blind review by at least two members of the International Program Committee. AIMSA 2024 attracted high-quality papers by researchers from 17 countries: Bulgaria, UAE, Ireland, China, Lebanon, Portugal, Iraq, Germany, Japan, Slovenia, India, Norway, Switzerland, Romania, Greece, New Zealand, and the UK.

The conference attracted as plenary speakers 3 leading academics and researchers in the area of AI who presented exciting presentations covering advanced AI methods and techniques.

- Prof. Plamen Angelov, from Lancaster University, UK, presented novel methods for explainable AI.
- Prof. Zeng-Guang Hou, from Chinese Academy of Sciences Institute of Automation in Beijing, China presented advanced robotic applications of AI in health and medicine.
- Prof. Giacomo Indiveri, Director of the Institute for Neuroinformatics (INI), from the University of Zurich and ETH, Switzerland, presented a compelling talk on why the next AI systems should be based on low-energy and high-accuracy neuromorphic platforms, both digital and analogue.

The post-conference proceedings include papers covering a variety of aspects of AI and its applications: natural language processing, sentiment analysis, image processing, optimization, reinforcement learning, from deep ANNs to spike timing NNs, and applications in economics, medicine, and process control.

We would like to thank all participants, authors, and plenary speakers, for making this conference a high-quality and a very inspirational one!

September 2024
Nikola Kasabov
Petia Koprinkova-Hristova

Organization

Programme Committee Chair

Nikola Kasabov   Auckland University of Technology, New Zealand
                                          Bulgarian Academy of Sciences, IICT, Bulgaria
                                          Ulster University, UK
                                          Dalian University, China

Local Organizing Committee Chairs

Petia Koprinkova-Hristova	Bulgarian Academy of Sciences, IICT, Bulgaria
Gennady Agre	Bulgarian Academy of Sciences, IICT, Bulgaria
Ivo Marinchev	Bulgarian Academy of Sciences, IICT, Bulgaria
Ivelina Nikolova	Bulgarian Academy of Sciences, IICT, Bulgaria

International Programme Committee

Alan Wang	University of Auckland, New Zealand
Albena Tchamova	Bulgarian Academy of Sciences, IICT, Bulgaria
Álvaro Huertas García	Polytechnic University of Madrid, Spain
Amedeo Napoli	LORIA, France
Aysegul Ucar	Firat University, Turkey
Banu Diri	Yildiz Technical University, Turkey
Bulent Bolat	Yildiz Technical University, Turkey
Catalin Stoean	University of Craiova, Romania
Claudio Gallicchio	University of Pisa, Italy
Costin Badica	University of Craiova, Romania
Damien Coyle	Ulster University, UK
Dan Tufis	RACAI, Romania
Diego Calvanese	Free University of Bozen-Bolzano, Italy
Dominik Slezak	University of Warsaw, Poland
Drago Žagar	University of Osijek, Croatia
Francesco Carlo Morabito	UNIRC, Italy
Galia Angelova	Bulgarian Academy of Sciences, IICT, Bulgaria
Genci Capi	Hosei University, Japan

Gennady Agre	Bulgarian Academy of Sciences, IICT, Bulgaria
George D. Magoulas	Birkbeck, University of London, UK
Christos Makris	University of Patras, Greece
Florin Leon	Gheorghe Asachi Technical University of Iasi, Romania
Helena Bahrami	Auckland University of Technology, New Zealand
Horia Pop	Babeş-Bolyai University, Romania
Iman Abouhassan	Technical University of Sofia, Bulgaria
Ioan Sacala	Universitatea Politehnica din București, Romania
Ireneusz Czarnowski	Gdynia Maritime University, Poland
Jie Yang	Shanghai Jiao Tong University, China
Jolanta Mizera-Pietraszko	Military University of Land Forces, Poland
Julita Vassileva	University of Saskatchewan, Canada
Kiril Simov	Bulgarian Academy of Sciences, IICT, Bulgaria
Konstantin Markov	University of Aizu, Japan
Konstantinos Demertzis	Democritus University of Thrace, Greece
Konstantinos Votis	CERTH-ITI, Greece
Krassimir Atanassov	IBBE, Bulgarian Academy of Sciences, Bulgaria
Loris Bozzato	Fondazione Bruno Kessler, DKM, Italy
Maciej Kandula	J&J Innovative Medicine, Austria
Marcin Paprzycki	SRI PAS, Poland
Mihai Gabroveanu	University of Craiova, Romania
Maria Nisheva-Pavlova	Sofia University St. Kliment Ohridski, Bulgaria
Mauro Gaggero	National Research Council of Italy, Italy
Mirel Cosulschi	University of Craiova, Romania
Mirjana Ivanovic	University of Novi Sad, Serbia
Nikola Kasabov	Auckland University of Technology, New Zealand
Okyay Kaynak	Bogazici University, Turkey
Ozcan Kalenderli	Istanbul Technical University, Turkey
Panayiotis Vlamos	Ionian University, Greece
Petia Georgieva	University of Aveiro, Portugal
Petia Koprinkova-Hristova	Bulgarian Academy of Sciences, IICT, Bulgaria
Petko Valtchev	Université du Québec à Montréal, Canada
Petr Hajek	University of Pardubice, Czech Republic
Richard Chbeir	Université de Pau et des Pays de l'Adour, France
Roman Barták	Charles University, Czechia
Roumen Trifonov	Technical University of Sofia, Bulgaria
Ruggero Donida	Università degli Studi di Milano, Italy
Seiichi Ozawa	Kobe University, Japan
Shihua Zhou	Dalian University, China
Soon Chun	City University of New York, USA
Stefan Trausan-Matu	University Politehnica of Bucarest, Romania

Stefka Fidanova	Bulgarian Academy of Sciences, IICT, Bulgaria
Stoyan Mihov	Bulgarian Academy of Sciences, IICT, Bulgaria
Tomas Krilavicius	Vytautas Magnus University, Lithuania
Tulay Yildirim	Yildiz Technical University, Turkey
Vladimir Kurbalija	University of Novi Sad, Serbia
Veselka Boeva	Blekinge Institute of Technology, Sweden
Vincenzo Piuri	Università degli Studi di Milano, Italy
Yancho Todorov	VTT Technical Research Centre of Finland, Finland

Contents

Multimodal Sentiment Analysis: Recognizing Sentiment in Memes 1
   Georgi Vankov, Dimitar Dimitrov, Ivan Koychev, and Preslav Nakov

Remote Sensing Data for Predicting Crop Growth 12
   Yunan Li, Sahraoui Dhelim, Liming Chen, and M-Tahar Kechadi

Cross-Lingual Style Transfer TTS for High-Quality Machine Dubbing 27
   Stoyan Mihov

An Approach to Discovering, Tracking Over Time, and Summarizing
Publicly Available Information on a Given Topic 40
   Alexandrina Karakehayova and Maria Nisheva-Pavlova

Reinforcement Learning Control of Cart Pole System with Spike Timing
Neural Network Actor-Critic Architecture 54
   Borislav Markov and Petia Koprinkova-Hristova

Predictive and Explainable Modelling in Economics on the Case Study
of Remittance Prediction Using the NeuDen AI Computational Architecture ... 64
   Iman AbouHassan, Nikola Kasabov, Roumen Trifonov, and George Popov

Deep Learning for Multi-class Diagnosis of Thyroid Disorders Using
Selective Features 80
   Filipa Santana, José Brito, and Petia Georgieva

Advanced CNN-SVM Machine Learning Techniques for Facial Skin
Ultrasound Image Analysis 93
   *Aayad Nabeel, Mostafa Ragheb, Galina Momcheva, Issa Kamar,
   and Mohamad Hamady*

Testing the NEAT Algorithm on a PSPACE-Complete Problem 104
   Angel Marchev Jr., Dimitar Lyubchev, and Nikolay Penchev

Investigating the Regularization of Deep Neural Networks for Affect
Recognition with Relevance-Guided Local Explanations 122
   Ines Rieger

Layered Data-Centric AI to Streamline Data Quality Practices
for Enhanced Automation 128
   *Muhammad Uzair Akmal, Saara Asif, Leonid Koval,
Selvine G. Mathias, Simon Knollmeyer, and Daniel Grossmann*

Combining Graph NN and LLM for Improved Text-Based Emotion
Recognition 143
   Xinhao Zou and Konstantin Markov

A Novel Study on Modelling and Adaptive Optimal Control of a Tubular
Reactor Based on Gaussian Processes 155
   Alexandra Grancharova, Junhong Xie, and Juš Kocijan

Converging Dimensions: Information Extraction and Summarization
Through Multisource, Multimodal, and Multilingual Fusion 168
   *Pranav Janjani, Mayank Palan, Sarvesh Shirude, Ninad Shegokar,
and Faruk Kazi*

Enhancing Question Answering in Lecture Videos with a Multimodal
Retrieval-Augmented Generation Framework 184
   Thomas Tanner, Andreas Marfurt, and Hasan Oğul

Agent-Based Simulation Leveraging Declarative Modeling for Efficient
Resource Allocation in Emergency Scenarios 199
   *Ionuţ Murareţu, Alexandra Vultureanu-Albişi, Sorin Ilie,
and Costin Bădică*

Enhancing Security in Federated Learning: Detection of Synchronized
Data Poisoning Attacks 211
   Dimitrios Anastasiadis and Ioannis Refanidis

Clinical and Acquisition Data for Optimizing MGMT Methylation
Status Prediction: A Comprehensive Ensemble Strategy Emphasizing
Non-invasive Approaches 223
   Mariya Miteva and Maria Nisheva-Pavlova

Author Index 237

Multimodal Sentiment Analysis: Recognizing Sentiment in Memes

Georgi Vankov[1](✉)[ID], Dimitar Dimitrov[1](✉)[ID], Ivan Koychev[1](✉)[ID], and Preslav Nakov[2](✉)[ID]

[1] Faculty of Mathematics and Informatics, Sofia University "St. Kliment Ohridski", Sofia, Bulgaria
`ggb9898@gmail.com`, `{ilijanovd,koychev}@fmi.uni-sofia.bg`
[2] Mohamed bin Zayed University of Artificial Intelligence, Abu Dhabi, United Arab Emirates
`Preslav.Nakov@mbzuai.ac.ae`

Abstract. The usage of memes and other visual material coupled with text on social media has been on the rise recently. Recognizing that visual signals are consumed quickly and can trigger emotional responses. It has become essential to discern the sentiment of such content, as it could significantly influence social media users. The paper focuses on the sentiment of memes on popular social networking platforms such as Instagram, Reddit, Facebook, and Tumblr. Our goal is to understand how these memes affect people in a positive, negative, or neutral way. We create a balanced dataset of 5,592 memes using distant supervision, i.e., automatically assigning sentiment labels based on different social media attributes, e.g., hashtags. We verify the accuracy of these labels by manually checking a random subset of the data. We conduct unimodal and multimodal experiments to explore how different cues contribute to identifying sentiment. Our results show that multimodal approaches, combining images and text, effectively identify the emotions in memes. We further experiment with novel closed and open-source LLMs, and we show that they outperform traditional multimodal approaches. The dataset is released publicly.

Keywords: Multimodal Sentiment Analysis · Emotion recognition · Meme dataset

1 Introduction

Memes have become an integral part of the social media landscape [2,18]. These humorous illustrations are widely shared on social platforms and offer an entertaining way to express personal perspectives or to react to situations online. However, memes are not just a source of digital giggles. They are a treasure trove of information about the collective sentiments of online communities. Sentiment analysis is a key tool that aids in understanding different media [12]. By studying memes through sentiment analysis, we explore the overall sentiment within various online platforms [22]. This understanding is valuable as it gives us insights into how memes are shared and spread in the digital world, how humour is used in conveying sentiments, and how to improve digital communication strategies [11]. However, memes pose a unique challenge for sentiment analysis. As seen in Fig. 1 the combination of visuals and text can make the sentiment conveyed more complex to decipher. That is why our study aims to explore the sentiment expressed by memes, considering challenges like bullying and hate speech [10] that may emerge in this online space.

Recently, methods of communication have grown towards visual expression rather than just textual communication between users. With that, social media platforms have adapted by introducing different types of metadata, e.g., reactions, hashtags, and focused topics. These attributes form clusters of similar content, which is why we leverage them when creating our dataset, which addresses one of the main challenges in multimodal sentiment analysis: the small amount of available data. We construct a multimodal dataset of memes in English by automatically assigning labels based on different types of attributes present in social networks, like hashtags. We verify the automatic annotations by manually checking a random sub-sample of 200 memes, where we observe 89% similarity between the automatic and manual labels, solidifying the initial assumption that different metadata attributes are a reliable source of information about meme sentiment.

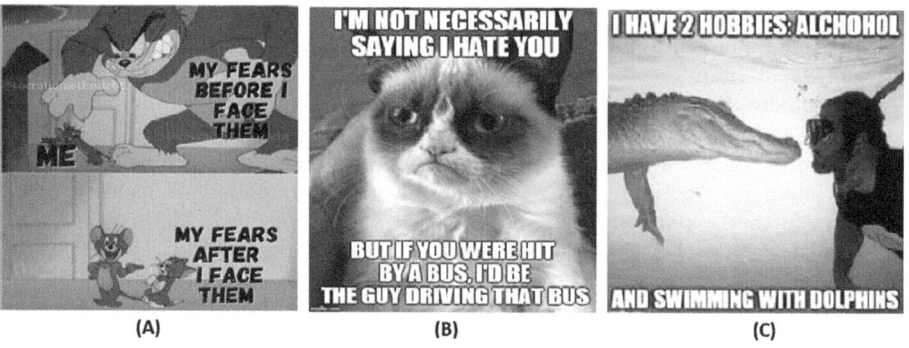

Fig. 1. Examples of memes with annotated sentiment: positive (A), negative (B), and neutral (C).

Furthermore, we conduct a series of unimodal and multimodal experiments that reveal that text or image alone is not enough to comprehend the sentiment inside a meme completely; instead, combining vision and language modalities is inherently better. We also show that traditional multimodal techniques such as CLIP and VILT are outperformed by the recent advances in closed and open-source LLMs such as GPT-3.5, GPT-4, and Llava.

Our contributions can be summarized as follows:

- We create a new dataset using distant supervision, consisting of 5,592 memes, which are annotated with three labels: positive, negative and neutral.[1] The dataset is of high quality: our manual analysis shows an 89% agreement rate between automatic and manual labels for a random subset of the data.
- We conduct experiments using state-of-the-art textual, visual, and multimodal models, emphasizing the significance of both modalities in understanding sentiments conveyed by memes.
- We further experiment with novel closed and open-source LLMs, showing that they outperform traditional multimodal approaches.

2 Related Work

Extensive research has been conducted in the field of meme sentiment analysis. Notably, the findings of Memotion 2, as summarized by [13], provide valuable insights into the latest techniques used to analyze emotions in memes. Participating teams utilized advanced deep learning models, such as EfficientNet-v2, to analyze image features, while models like BERT or RoBERTa were employed to process textual content. Combining image and text analysis has shown promising results, highlighting the importance of considering multiple aspects to capture the nuanced emotions conveyed in memes. In contrast to the Memotion study, our research offers two significant advancements. Firstly, we utilize an up-to-date dataset collected from diverse social media platforms, ensuring a comprehensive representation of online sentiments in memes. Secondly, we employ state-of-the-art models such as GPT-4 [7] and VILT [8], which integrate advanced natural language understanding and visual processing techniques.

Ferrara, E. et al. [4] conducted significant research on meme analysis and clustering in social media. They focused on meme detection by clustering messages from various large social data streams, specifically Twitter. The dataset used in their study was primarily limited to textual elements, such as messages, tweets, and trending hashtags. In contrast, our research expands upon this by incorporating images and text, providing a more comprehensive analysis of memes.

In modern memes, characterized by images accompanied by embedded text, extracting features from both modalities during meme sentiment analysis becomes essential.

While previous studies [12,13] have contributed to multimodal sentimental analysis, they exhibit specific weaknesses and limitations. For instance, [6] focused on experiments using Tumblr posts that do not precisely represent memes, as memes have texts embedded within the images, whereas Tumblr posts have separate captions. Similarly, [15] solely performed binary classification to detect offensive memes without providing a comprehensive analysis of the generic sentiments expressed in internet memes. Recent studies have attempted to prompt Pretrained Language Models (PLM) and yield good performance for uni-modal NLP [1,5,16,17]. Nevertheless, few works have attempted to prompt PLMs for multimodal tasks [5,21]. [20] has explored prompting GPT-4 model

---

[1] Dataset will be released publicly after paper acceptance.

[1] for the visual question and answering task. However, the approach has limitations as large models such as GPT-3 are expensive to tune. In contrast, our study leverages the advanced GPT-4 model for prompt-based analysis, providing improved capabilities compared to the GPT-3-based solution proposed by [1].

Despite the progress made in meme sentiment analysis, the fast-paced nature of memes and their extensive use of cultural references continue to present challenges. Our research aims to contribute to this field by exploring innovative approaches and leveraging updated datasets to enhance the accuracy and effectiveness of sentiment analysis in memes.

3 Dataset Creation

In this section, we describe the construction of our dataset.

3.1 Data Collection

We employed a combination of APIs and web scraping techniques to collect the dataset to extract meme images from popular social networking platforms, including Instagram, Reddit, Facebook, and Tumblr. Our focus was on gathering a diverse range of meme images that represent different sentiments.

Instagram: We utilized the Instagram API to extract images, with hashtags such as **#negative_vibe_memes** for negative memes and **#make_me_smile** for positive memes

Reddit: The Reddit API allowed us to collect meme content from relevant subreddits, such as **r/depression_memes** for negative and **r/goodvibememes** for positive memes

Facebook: With the help of Apify, we sourced publicly accessible meme content. We utilized groups such as **"Daily positive memes"** to extract positive memes and **"Depression Memes Posting"** for negative memes. This strategic approach allowed us to ensure a well-rounded representation of sentiments in our dataset.

Tumblr: We use the Tumblr API and provided hashtags like: **#dankmemes** to extract negative memes and **#positive_vibe** for positive memes.

The dataset annotation was done fully automatically, where we assigned sentiment classes (positive, negative, and neutral) to the collected images based on relevant hashtags from the different sources, enabling the categorization of memes by their predominant sentiment. We filtered out non-English memes and duplicates to ensure dataset quality and uniqueness. We also used OCR[2] technology to extract textual content from the memes, enriching the dataset for comprehensive multimodal analysis.

We randomly selected 200 memes from our dataset to validate our automated approach and manually labelled them based on sentiment. The result showed an

---

[2] pytesseract - Python wrapper for Google's Tesseract-OCR Engine.

89% similarity between automatic and manual labelling, suggesting a high level of reliability in our automated sentiment categorization. This validation process gave us confidence in our methodology and the resulting dataset, paving the way for an in-depth exploration of sentiments conveyed through the colourful world of memes.

3.2 Data Preprocessing

Once the dataset was collected, we performed the following preprocessing steps to prepare it for further analysis:

Image Resizing: To ensure uniformity in the dataset, we resized all the images to a consistent resolution (224 × 224). This step eliminates discrepancies in image dimensions, enabling a more reliable comparison and analysis.

Text Normalization: We use text normalization techniques to standardize the textual content within the memes. This process includes converting text to lowercase, removing special characters, and addressing format inconsistencies.

Noise Removal: To enhance the overall quality of the dataset, we removed any unnecessary noise or artefacts present in the images. This involved tasks such as eliminating image blurriness, reducing compression artefacts, and addressing any anomalies that could affect the interpretability of the images.

These steps collectively ensure that the dataset is well-structured, consistent, and free from extraneous elements, setting the foundation for accurate and meaningful analysis.

3.3 Dataset Statistics

Our dataset is evenly distributed across the three sentiment classes: positive, negative, and neutral. This equal distribution ensures that our models are trained and evaluated on diverse meme images, capturing the full spectrum of sentiments. The dataset consists of a total of 5,592 meme images. Table 1 presents the distribution of the collected data into different sources and classes.

Table 1. Statistics about the datasets.

	Positive	Negative	Neutral	Total
Instagram	630(32.2%)	671(34.3%)	654(33.5%)	1955(35%)
Tumblr	384(27.3%)	432(30.7%)	589(41.9%)	1405(25.1%)
Reddit	525(37.7%)	478(34.4%)	387(27.9%)	1390(24.9%)
Facebook	319(37.9%)	299(35.5%)	224(26.6%)	842(15%)
Total	1858(33.23%)	1880(33.63%)	1854(33.14%)	5592(100%)

To evaluate the performance of our models, we divided the dataset into train and test sets. The train set, constituting 70% of the data, was used for model training and parameter optimization. The remaining 30% of the data was reserved for testing the generalization and effectiveness of the trained models.

4 Experiments and Evaluation

We use standard classification metrics: Accuracy, Precision, Recall and F1_score. Each experiment was repeated three times for reliability, providing an average result. Experiments were performed using APIs or Nvidia A100 GPU.

In this section, we present the models employed for sentiment analysis of memes, considering different approaches that utilize text-only, image-only, and multimodal (text and image) techniques. We explore the following models: BERT, VILT [8], CLIP (ViT-B/32), CLIP (ViT-L/14), VGG19 [19], LLaVA [9] and GPT-4 [7] with Prompting.

The selection of these models is driven by their distinct capabilities in handling different aspects of meme content. BERT excels in text understanding, VILT integrates vision and language data, CLIP demonstrates powerful image-text alignment, VGG19 is a proven image classification architecture, LLaVA specializes in visual-linguistic tasks, and GPT-4 with Prompting combines contextual understanding of both text and image. This diverse set allows us to comprehensively evaluate and compare performance across various sentiment analysis methodologies for memes.

4.1 Text-Only Experiments

BERT [3]: We use BERT to process the textual content. The extracted text is fed into the base BERT model for Sequence Classification.

CLIP (Text-Only): [14] We also utilize the CLIP model with only the textual component. First, the CLIP model processes the textual content from memes to create encoded text representations. These representations are then fed into the Logistic Regression classifier for sentiment analysis.

4.2 Image-Only Experiments

VGG19 [19] We leveraged the VGG19 architecture to encode the visual features of memes, enabling us to conduct sentiment analysis relying solely on visual cues.

CLIP (Image-Only) [14] Both versions of a CLIP model were used to encode the visual features of meme images, allowing us to perform sentiment analysis based on visual cues alone.

4.3 Multimodal Experiments

In the multimodal approach, we combine memes' textual and visual information. We utilize the following models and techniques:

VILT [19] The VILT model is employed for multimodal sentiment analysis. It enables the fusion of the textual content with the visual features of the images, capturing the combined sentiment expressed through text and image.

CLIP (Early Fusion) [14] In this approach, we utilize the CLIP model with early fusion. The encoded image and text representations are concatenated before being provided to the classifier.

GPT-4 and GPT-3.5 with Prompting [7] To incorporate the contextual understanding of both text and image, we utilize GPT-4 and GPT-3.5 with Prompting. In Fig. 2, we show an example prompt. The expected input is the textual content of a meme and the image caption[3]. The desired output is one of the categories: positive, negative, or neutral.

Table 2. Experimental results using text-only, image-only, and text+image models. Models that use both text and image data outperform text-only and image-only approaches.

	Model	Accuracy	Precision	Recall	F1
	Majority class(Baseline)	0.35	0.12	0.33	0.17
Text only	BERT	0.58	0.60	0.57	0.57
	CLIP (ViT-B/32)	0.59	0.61	0.57	0.58
	CLIP (ViT-L/14)	0.60	0.60	0.60	0.60
Image only	VGG19	0.57	0.59	0.66	0.58
	CLIP (ViT-B/32)	0.68	0.68	0.68	0.68
	CLIP (ViT-L/14)	0.74	0.74	0.74	0.74
Text + Image	GPT-4v Preview	0.63	0.64	0.62	0.62
	VILT	0.71	0.73	0.69	0.70
	CLIP (ViT-B/32)	0.69	0.69	0.69	0.69
	CLIP (ViT-L/14)	0.75	0.75	0.75	0.75
	GPT-3.5 Prompt	0.74	0.75	0.76	0.75
	GPT-4 Prompt	0.77	0.79	0.79	0.80
	LLaVA	**0.79**	**0.81**	**0.81**	**0.82**

"Please classify the sentiment of the following meme's text and caption into one of the categories: positive, negative, or neutral.

Meme Text: 'when you convince your friend who'd rather stay home to hang out.'

Meme Caption: 'The image features a smiled brown dog sitting next to a sad-looking cat.'

Sentiment Category: [Select one: Positive, Negative, Neutral]

Answer: Positive"

Fig. 2. Our GPT-4 text-only prompt.

---
[3] For generating image caption, we use CLIP.

LLaVA [9] In the LLAVA prompt, we've simplified the process by directly using the image without needing a textual description. In Fig. 3, we show an example prompt. The expected input is the raw image and the question, "Classify the sentiment of the provided meme into one of the following categories: positive, negative, or neutral.". The desired output is one of the categories: positive, negative, or neutral.

Fig. 3. An example of LLaVA prompt.

GPT-4 Vision Preview. In our experiment with the GPT-4 Vision Preview, we've adapted the successful approach used in the LLAVA model. The experiment involved directly inputting images into the model without relying on textual interpretations. The same sentiment classification question used in LLAVA was also employed here. This allowed us to gauge GPT-4's ability to interpret and classify image-based sentiments, highlighting its developments in understanding visual data.

5 Comparison of Models

In this section, we present the evaluation metrics and model comparison results for our sentiment analysis task. We compare the performance of different models, including text-only, image-only, and text+image approaches.

Table 2 showcases the model evaluation results. Each row in the table represents a specific model, while the columns correspond to the evaluation metrics. We use a macro-level evaluation approach, considering the overall performance across all sentiment categories.

Overall, the multimodal models, combining text and image data, outperform the text-only and image-only approaches. Considering both modalities, the models capture more comprehensive information and achieve higher accuracy.

Among the models tested, the open-source LLaVA Prompting model performed best, providing accurate results due to its advanced language generation and ability to work with raw images. GPT-4 also showed promising results, considering the simple prompt strategy used. The CLIP models, which utilized both textual and visual inputs, also performed well, highlighting the importance of leveraging both modalities for improved analysis.

6 Conclusion

We have constructed a new multimodal dataset featuring 5,592 memes, categorized into positive, negative, or neutral emotions using distant supervision. Our data compilation spans various social media platforms, including Instagram, Reddit, Facebook, and Tumblr. We underscore the significance of considering textual and visual elements in meme sentiment analysis through unimodal and multimodal experiments.

Our investigations extended to GPT-3.5, GPT-4, and LLaVA prompting, showcasing substantial performance superiority over other methods and achieving an impressive F1 Score of 0.82. Significantly, we contribute to the field by creating a robust dataset that can serve as a valuable resource for training and evaluating the capabilities of new large language models.

In future work, we plan to expand our meme collection further, incorporating memes from various languages. Our primary focus remains on providing a powerful training ground for emerging language and vision models, enhancing their ability to comprehend the intricate emotions conveyed by memes through the synergistic interplay of text and images. This research sets the stage for advancements in meme sentiment analysis by facilitating the development and evaluation of state-of-the-art language models.

Acknowledgement. This work is partially funded by the EU through NextGenerationEU funds, through the National Recovery and Resilience Plan of the Republic of Bulgaria, project No BG-RRP-2.004-0008.

References

1. Brown, T.B., et al.: Language models are few-shot learners. In: Advances in Neural Information Processing Systems: Annual Conference on Neural Information Processing Systems, NeurIPS (2020)
2. Chen, G.M., Zhang, D., Yang, S.: Memes in a digital world: recontextualization, creativity, and civic participation. Soc. Media + Soc. **6**(4), 2056305120970803 (2020)
3. Devlin, J., Chang, M.-W., Lee, K., Toutanova, K.: Bert: pre-training of deep bidirectional transformers for language understanding. arXiv preprint arXiv:1810.04805 (2018)

4. Ferrara, E., JafariAsbagh, M., Varol, O., Qazvinian, V., Menczer, F., Flammini, A.: Clustering memes in social media. In: 2013 IEEE/ACM International Conference on Advances in Social Networks Analysis and Mining (ASONAM 2013), pp. 548–555. IEEE (2013)
5. Gao, T., Fisch, A., Chen, D.: Making pre-trained language models better few-shot learners. In: Proceedings of the Annual Meeting of the Association for Computational Linguistics and the International Joint Conference on Natural Language Processing, ACL/IJCNLP, pp. 3816–3830 (2021)
6. Hu, A., Flaxman, S.: Multimodal sentiment analysis to explore the structure of emotions. In: Proceedings of the 24th ACM SIGKDD International Conference on Knowledge Discovery & Data Mining, pp. 350–358 (2018)
7. Katz, D.M., Bommarito, M.J., Gao, S., Arredondo, P.: GPT-4 passes the bar exam. arXiv preprint arXiv:2212.14402 (2023). Available at SSRN: https://ssrn.com/abstract=4389233. https://doi.org/10.2139/ssrn.4389233
8. Kim, W., Son, B., Kim, I.: ViLT: vision-and-language transformer without convolution or region supervision. In: International Conference on Machine Learning (2021)
9. Liu, H., Li, C., Wu, Q., Lee, Y.J.: Visual instruction tuning. arXiv preprint arXiv:2304.08485 (2023)
10. Maity, K., Jha, P., Saha, S., Bhattacharyya, P.: A multitask framework for sentiment, emotion and sarcasm aware cyberbullying detection from multi-modal code-mixed memes. In: Proceedings of the 45th International ACM SIGIR Conference on Research and Development in Information Retrieval, pp. 1739–1749 (2022)
11. Nascimento, L., Oliveira, L., Pereira, A., Moro, M.M.: Exploring emotions and social influence in online social networks. J. Inf. Sci. **46**(4), 515–532 (2020)
12. Pang, B., Lee, L.: Opinion mining and sentiment analysis. Found. Trends Inf. Retr. **2**(1–2), 1–135 (2008)
13. Patwa, P., et al.: Findings of memotion 2: sentiment and emotion analysis of memes. In: Proceedings of De-Factify: Workshop on Multimodal Fact Checking and Hate Speech Detection (2022)
14. Radford, A., et al.: Learning transferable visual models from natural language supervision. In: International Conference on Machine Learning (2021)
15. Sabat, B.O., Ferrer, C.C., Giro-i Nieto, X.: Hate speech in pixels: detection of offensive memes towards automatic moderation. arXiv preprint arXiv:1910.02334 (2019)
16. Schick, T., Schütze, H.: Exploiting cloze-questions for few-shot text classification and natural language inference. In: Proceedings of the Conference of the European Chapter of the Association for Computational Linguistics: Main Volume, pp. 255–269 (2021)
17. Schick, T., Schütze, H.: It's not just size that matters: small language models are also few-shot learners. In: Proceedings of the Conference of the North American Chapter of the Association for Computational Linguistics: Human Language Technologies, NAACL-HLT, pp. 2339–2352 (2021)
18. Shifman, L.: Memes in Digital Culture. MIT Press (2014)
19. Sudha, V., Ganeshbabu, D.: A convolutional neural network classifier VGG-19 architecture for lesion detection and grading in diabetic retinopathy based on deep learning. Comput. Mater. Continua **66**, 827–842 (2020)

20. Yang, Z., et al.: An empirical study of GPT-3 for few-shot knowledge-based VQA. CoRR (2021)
21. Yao, Y., Zhang, A., Zhang, Z., Liu, Z., Chua, T.-S., Sun, M.: CPT: colorful prompt tuning for pre-trained vision-language models. CoRR (2021)
22. Yoon, J., Song, Y.: A study on the effectiveness of using social media data for sentiment analysis. IEEE Access **9**, 13448–13456 (2021)

Remote Sensing Data for Predicting Crop Growth

Yunan Li[1(✉)], Sahraoui Dhelim[1], Liming Chen[2], and M-Tahar Kechadi[1]

[1] School of Computer Science, University College Dublin (UCD), Belfield, Dublin 4, Dublin, Ireland
`Yunan.li@ucdconnect.ie`, {`sahraoui.dhelim,tahar.kechadi`}`@ucd.ie`
[2] School of Computer Science and Technology, Dalian University of Technology, Dalian, China
`limingchen0922@dlut.edu.cn`

Abstract. This paper introduces an efficient approach, the dynamic coefficient polynomial model, which emulates crop growth dynamics using NDVI. This model, a significant improvement over traditional models like the NDVI mean and static polynomial models, is designed to be adaptable over time and incorporates spatial variables to account for the diverse growth conditions experienced in different regions. Consequently, the model's responses and adaptations are influenced by the specific crop growth dynamics observed within these spatial dimensions, adding a new dimension to crop growth forecasting. Our results show that the proposed model achieves a higher accuracy than the other machine learning models, which is about 90.3%.

Keywords: Sentinel-2 · NDVI · Moore-Penrose inverse · Machine Learning · Deep Leaning

1 Introduction

Remote sensing data has become increasingly important for various applications in agriculture, including delineating farmland imagery, monitoring soil dynamics, and assessing crop development. Studies have demonstrated the effectiveness of remote sensing techniques in crop classification [1], cropland mapping [2], and crop growth monitoring [3].

The Normalized Difference Vegetation Index (NDVI) is a commonly used metric in agriculture and remote sensing to measure vegetation health and density. It analyzes the difference between near-infrared and red-light reflectance from vegetation, providing insights into plant vigor, biomass production, and overall crop health. High NDVI values indicate dense, healthy vegetation, while low values may indicate stress or sparse vegetation cover. Considering the stable growth cycle of a crop, NDVI is the preferred sensory data for charting its developmental phases. Gaussian functions have been used to determine the growth trends of crops and their spatial distribution [4]. Moreover, interesting approaches using NDVI-derived parameters have been developed to explore the geographical extent of crop cultivation [5].

Predicting crop growth stages is crucial for optimizing agricultural management operations and maximizing crop yields. This task involves assessing various factors, such as temperature, moisture levels, and photoperiod, which influence the developmental progression of crops from planting to maturity. Traditional methods rely on phenological observations and accumulated degree days, but advancements in remote sensing, particularly utilizing indices like NDVI, offer opportunities for more accurate and timely predictions. Integrating satellite imagery and machine learning algorithms allows for the monitoring of crop growth dynamics on a large scale, enabling farmers to make informed decisions regarding irrigation, fertilization, and pest management. However, challenges persist, including the need for ground-truth data and accounting for environmental variability. Nonetheless, ongoing research and technological innovations continue to improve the reliability and efficiency of predicting crop growth stages using sensor data, benefiting agricultural productivity and sustainability.

This paper focuses on predicting crop growth, particularly winter wheat. While previous works [6] and [7] have outlined methodologies for forecasting regional growth trends based on NDVI, they rely heavily on experts' knowledge, making them less efficient when applied to new geographic regions. Our approach uses big data analytics on very large arable areas with different characteristics, therefore dealing with key limitations of the existing approaches.

2 Related Work

Numerous research works have studied the positive correlation between NDVI and vegetation growth [8]. These studies involve a systematic delineation of vegetation growth stages using the BBCH[1] table, followed by an evaluation of the potential applicability of NDVI for predictive crop growth stages. In [9], the authors focused on distinct vegetation indices, including NDVI, NDBR (Normalized Burn Ratio), and NDMI (Normalized Difference Moisture Index) [10]. Their goal is to elucidate vegetation dynamics through the utilization of broad-spectrum information. NDVI performed better than other indices, withg correlation coefficients (R^2) ranging from 0.6 to 0.99 across different phases. Similarly, [11] investigated and explored the correlation between the BBCH scale and observed NDVI, revealing significant correlation coefficients (R^2 or coefficient of determination) of 0.93 for winter wheat and 0.77 for oilseed rape. Moreover, [12] developed 19 growth metrics based on NDVI and Enhanced Vegetation Index (EVI), including Vegetation Growth Metrics (VGM), Maximum (VGMmax), and Integrated (VGMinteg). A subset of these variables exhibited high R^2 values of approximately 95%, reaffirming the strong correlation between NDVI and plant growth.

---

[1] Biologische Bundesanstalt, Bundessortenamt und Chemische Industrie.

We identified three main categories of methodologies. 1) The first category uses only NDVI data and derives the range of measures such as average, maximum, and minimum values [6]. 2) The second category creates new variables like dNDVI and pNDVI, as explained in [7]. 3) The third category, [13], uses a polynomial function.

It has been observed that crop growth can be predicted accurately by using only NDVI data. In [6], a comprehensive investigation was carried out, focusing on NDVI time series datasets within the Samara Airport Area, Russia, 2018. The study classified crops into different stages using NDVI ranges (Mean, MAX, Min, and standard deviation). This method achieved 92% accuracy, demonstrating the potential for accurate monitoring of crop phenology at various growth stages. While some stages used simplistic Mean, Max, and Min methods, the proposed dynamic coefficient prediction model returned higher accuracy in predicting specific NDVI values daily. This model is highly beneficial, especially in extreme weather events, where values deviate beyond the common range. Similarly, [14] used remote sensing data to predict soybean growth. They used two-time series-based baseline estimators: simple polynomial interpolation and a dual logistic model. The latter model, [13], relies on the maximum and minimum NDVI values in winter. However, this model cannot be used when ground-level crop growth data is unavailable. Furthermore, the study [14] suggested a third-order model, which, although inspiring, lacks the adaptability offered by the dynamic coefficient prediction model. The dynamic model returned optimal results with a sixth-order equation when we researched winter wheat, demonstrating adaptability to diverse growth scenarios.

Moreover, the proposed dynamic coefficient prediction model is designed to predict missing values caused by cloud cover. To address cloud-induced gaps, we use an approach detailed in [15]. This approach mainly employs the least squares method to fit vegetation indices by using a local polynomial function to determine the function's value during specific time periods. The Savitzky-Golay filter can be used in this case, but it is susceptible to residual noise contamination. Therefore, our model uses a broader fitting approach and direct noise filtration to mitigate potential influences on the obtained results.

In summary, the existing studies stress the importance of using advanced models, like the dynamic coefficient prediction, to forecast crop growth stages accurately. These models help address the limitations of traditional monitoring methods and promote adaptability to dynamically changing environmental conditions.

Machine learning and deep learning techniques are extensively utilized for yield prediction across various crops, such as maize [16], rice [17], and wheat [18]. These methods are also applied to predict different crop growth stages, such as corn [19] and soybean maturity [20]. In [21], they used Random Forest and Deep Learning models to predict and estimate the rice LAI. They used NDVI and climate monitoring to reproduce the LAI of rice. Depending on the location, they found that the model accuracy varies, with the best model efficiency being 0.88.

3 Dataset Collection and Preparation

We collected a dataset from 1732 arable fields located throughout the United Kingdom and Ireland between 2017 and 2020. The data source is mostly satellite imagery; Sentinel-2[2], an Earth observation mission from Copernicus Programme. Sentinel Hub offers various layers, such as NDVI, GDVI, EVI, NDRE, etc. While satellite images are easy to collect, they vary in quality and availability. For instance, they can be affected by clouds, snow, shadows, etc.

We identified a subset comprising 1557 fields with documented drilling date information. Furthermore, it is worth noting that some fields underwent winter wheat cultivation for many years, with certain fields hosting winter wheat crops for up to three consecutive years. In light of the variability in NDVI growth stages across diverse fields and years, each unique combination has been regarded as an individual entity. Consequently, our dataset contains 1,950 distinct groups of images corresponding to the 1,950 items gathered. Overall, the dataset encompasses 215,170 images, which is large enough for analysis.

The various indices mentioned above can also be calculated, given that we have the data for the 1313 spectral bands offered by Sentinel-2. The NDVI layer involves the NIR (B08) layer and the red (B04) layer, as expressed in Eq. (1).

$$NDVI = \frac{NIR - Red}{NIR + Red} = \frac{B08 - B04}{B08 + B04} \qquad (1)$$

We have opted for a resolution of (10 × 10) square meters for each field, wherein we procure GPS coordinates, NDVI, and all spectral bands. Leveraging the capabilities of the Sentinel Hub, we use the cloudy pixel identification model to assess the likelihood of cloud cover for each pixel.

4 Methodology

We first need to prepare the data before analyzing it. The data cleaning was conducted in two phases: initially, thresholds for cloudiness were determined, followed by removing extraneous noise artifacts. Identifying cloud cover thresholds is crucial, a topic that has not been extensively explored. Consequently, we introduce an efficient approach for selecting optimal cloudiness thresholds. We detail the theoretical foundations of our dynamic coefficient polynomial model and explain how it works. Furthermore, we developed a Deep Neural Network model to predict the distinct stages of six winter wheat species.

4.1 Data Cleaning

The data-cleaning process consists of four steps. First, we delineate the boundaries of the polygons to exclude pixels lying beyond these limits. Next, we identify a threshold for cloudiness to remove days with cloud cover, allowing us to isolate the cloudy segments. After that, we remove additional days affected by noise from the dataset.

---

[2] https://www.esa.int/Applications/Observing_the_Earth/Copernicus/Sentinel-2

4.1.1 Removing 24 Meters Boundaries and Beyond

We have the polygon boundaries of each field. However, upon collecting the dataset, we must specify the boundaries in a rectangular format, thus limiting the download to images contained within these rectangles; this has led us to download extra areas that are out of boundaries. This consists of creating new polygon boundaries, leaving the original boundaries at 24 m. Subsequently, we identify pixels falling within these newly defined boundaries using their GPS coordinates, retaining only this relevant portion of the dataset.

4.1.2 Identifying Cloudy Threshold

Sentinel 2 provides a method for identifying potential cloud cover by using information from its 13 spectral bands. However, it's difficult to determine the right threshold for identifying clouds. Lower thresholds for cloud probability result in clear images, but excluding all data with a cloud probability of zero leads to significant data loss, impacting research results due to the reduced dataset size. On the other hand, setting excessively high thresholds introduces a lot of noise into the dataset. It's important to strike a balance in defining the cloud threshold to ensure a dataset of sufficient size without compromising its integrity by excluding or including too much cloudy data.

We want to determine the best cloud cover threshold. NDVI indicates vegetation growth, but for winter wheat, once it reaches the flowering stages, NDVI keeps increasing. We select data from sowing to peak NDVI, assuming stability. Any further increase in NDVI is considered the ideal value for that timeframe. If NDVI falls below this mark, it's considered cloud-affected. We calculate the duration of these days but don't classify them as cloudy due to the "Unfounded Cloudy Days" threshold.

It's important to note that the adequacy of the threshold effect should not be solely evaluated based on the count of "Unfounded Cloudy Days." Each field and year have unique characteristics. Between 2016 and 2020, the frequency of Sentinel 2's image acquisition increased significantly from around 60 to 145 images per year. Areas with cloud possibilities exceeding the predefined threshold are designated as cloudy regions and are excluded from calculating the average daily NDVI values for the entire field. Additionally, if the entire field shows a high possibility of exceeding the threshold on a specific day, the corresponding daily NDVI values are omitted from the dataset, and these days are considered cloudy days.

As a result, the summary image numbers may vary when we use different thresholds. To solve this problem, we suggest using the "Unfounded Cloudy Days" percentage for all images of a specific field in a given year. This proposed method, as shown in Eq. (2), aims to evaluate the proportion of "Unfounded Cloudy Days" in comparison to the total observation period.

$$Percentage = \frac{1}{N} \sum_{n=0}^{N-1} \frac{\sum_j C_{\downarrow,j}}{\sum_i D_i} \qquad (2)$$

where $C_{\downarrow,j}$ represents the total count of the "Unfounded Cloudy Days", the $\sum D_i$ is defined as the summary of days after filter, i,j represents the ith and j^{th} entry in the date list.

4.1.3 Removing Other Noise

We filter out all regions affected by cloud cover, subsequently computing the Daily NDVI for each field. As illustrated in the case of Field ID 51258 during the year 2018, depicted in Fig. 1, a notable decrease in NDVI occurred on specific days, particularly towards the end of November and December. Such irregular fluctuations are atypical in practical scenarios. To investigate this anomaly, we observe the local weather conditions, focusing on the nearby weather station's snow cover data. Our examination reveals a correlation between the observed NDVI decreases and snowy conditions. While not all instances of snowfall lead to such drastic NDVI decreases, we attribute these two specific points to snow-related effects. Consequently, we classify all points that experienced a decrease to half their preceding value as instances of additional noise in the data. Removing such data from the dataset becomes imperative, as their inclusion could adversely affect predictive modeling by introducing confusion and distortion in the prediction dataset.

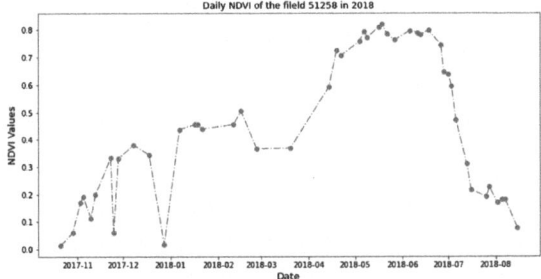

Fig. 1. The field 51258 in 2018 after removing cloudy area NDVI Daily value

4.2 Growth Curve Prediction

The model's objective revolves around examining the growth patterns of the crop. We systematically calculate the duration since the sowing date for each date in every field, denoting this variable as our input parameter. Simultaneously, we designate the NDVI value for each corresponding day as our output variable, representing our predictive target. Through the graphical representation of our findings, we observe a growth pattern resembling that of a K-order polynomial equation. As a result, we hypothesize the form of the prediction function illustrated in Eq. (3) to encapsulate this observed trend.

$$y = a_K \times x^K + a_{K-1} \times x^{K-1} + \cdots + a_1 \times x + a_0 \tag{3}$$

where y is the present NDVI value, and x is the number of days after the sowing date.

Given the uncertainty surrounding the precise K values within the function, this scenario presents an over-determined problem. Consequently, the determination of coefficients via the least-square estimation method becomes feasible and straightforward. For every field, a series of NDVI is denoted by $\boldsymbol{Y}$ (as indicated in Eq. (4)) alongside a corresponding list of Days, represented by $\boldsymbol{X}$ (as expressed in Eq. (5)). Furthermore, the coefficients matrix, designated as $\boldsymbol{A}$ (as illustrated in Eq. (6)), becomes a key element within this context.

$$\boldsymbol{Y} \triangleq \begin{bmatrix} y_1, y_2, \ldots, y_n, \ldots, y_N \end{bmatrix}^T \quad (4)$$

where the T represents the transpose of a matrix. The matrix $\boldsymbol{Y}$ is a one-column matrix with dimensions $N \times 1$.

$$\boldsymbol{X} \triangleq \begin{bmatrix} x_1^K, x_1^{K-1}, \ldots, x_1, 1 \\ \vdots, \vdots, \ddots, \vdots, \vdots \\ x_n^K, x_n^{K-1}, \ldots, x_n, 1 \\ \vdots, \vdots, \ddots, \vdots, \vdots \\ x_N^K, x_N^{K-1}, \ldots, x_N, 1 \end{bmatrix} \quad (5)$$

The matrix $\boldsymbol{X}$ is a matrix with dimensions $N \times (K+1)$.

$$\boldsymbol{A} \triangleq \begin{bmatrix} a_K, a_{K-1}, \ldots, a_1, a_0 \end{bmatrix}^T \quad (6)$$

The matrix $\boldsymbol{A}$ is a one-column matrix with dimensions $(K+1) \times 1$.

The formulation of our prediction function is represented by $\boldsymbol{Y} = \boldsymbol{X}\boldsymbol{A}$. When $\boldsymbol{X}$ takes the form of an $N \times N$ matrix, employing the least-square estimation facilitates the computation of coefficients $\boldsymbol{A}$, expressed as $\boldsymbol{A} = \boldsymbol{X}^{-1}\boldsymbol{Y}$. However, when $\boldsymbol{X}$ becomes an $N \times (K+1)$ matrix, it results in the nonexistence of an inverse matrix, $\boldsymbol{X}^{-1}$, leading us to seek an alternative solution. Hence, the necessity arises to determine the Moore-Penrose pseudo inverse of $\boldsymbol{X}$, $\boldsymbol{X}^{\dagger}$ denoted as depicted in Equation $\boldsymbol{X}^{\dagger} = (\boldsymbol{X}^T\boldsymbol{X})^{-1}\boldsymbol{X}^T$.

Then the least-square estimation facilitates the computation of coefficients $\boldsymbol{A}$ computed by $\boldsymbol{A} = \boldsymbol{X}^{\dagger}\boldsymbol{Y}$, where $\dagger$ represents the Moore-Penrose pseudoinverse.

After we compute the coefficients $\boldsymbol{A}$, we use Equation $\boldsymbol{Y}_{\text{prediction}} = \boldsymbol{X}\boldsymbol{A}$ to compute the prediction NDVI values, $\boldsymbol{Y}_{\text{prediction}}$.

NDVI Slope Computing: The Slope calculation for our NDVI is a direct derivative calculation using our dynamic polynomial model, calculated as follows Eq. (7):

$$y = K \times a_K \times x^{K-1} + (K-1) \times a_{K-1} \times x^{K-2} + \cdots + 2 \times a_2 \times x + a_1 \quad (7)$$

which $\boldsymbol{A}$ has already changed from $(K+1)$ elements to K elements, resulting in a dimension of $K \times 1$. Additionally, the matrix $\boldsymbol{X}$ has been adjusted to a dimension of $N \times K$.

4.3 Stages Prediction Model

The machine learning models utilized in this study are implemented using the `scikit-learn` package provided in Python3. The Deep Neural Network (DNN) model we designed was implemented using PyTorch. Given that our dataset comprises only two types, attempts to employ more complex DNN models resulted in overfitting issues, such as high training accuracy contrasted with low test accuracy. Consequently, the structure of the DNN model was simplified to include an input layer, ten hidden layers, and an output layer. The ReLU activation function was applied between the input and hidden layers, while the sigmoid function was used to connect the hidden layers to the output layer.

We created six different Deep Neural Network (DNN) models corresponding to six distinct stages, utilizing datasets specific to each stage. For instance, from a total of 600,000 data points, approximately 20,000 items pertain to the sowing stage. We randomly selected an additional 20,000 items from other stages, resulting in a dataset of 40,000 items used to generate model 1, which was then evaluated for accuracy. Note that 80% of the dataset was used for training, while the remaining 20% was reserved for testing. The accuracy was recorded for each model, and the average accuracy across all six models was computed as the final DNN result. The inputs are NDVI and NDVI slope, while the output is a prediction percentage. Predictions exceeding 0.5% were classified as the target stage, while those below were classified as other stages.

5 Experimental Results

As previously outlined, our data cleaning process encompasses several procedural steps. This section will elaborate on the methodology employed in parameter selection and subsequent model prediction outcomes. Specifically, we will delineate the rationale behind selecting the cloudy threshold and the optimal K value for the predictive models. Additionally, the predictive performance of three distinct baseline models and our dynamic polynomial model will be shown in figures and tables, compared against each other through RMSE metrics. In addition, we utilized the growth coefficients predicted for each field by the dynamic model to forecast the daily NDVI values and the slope of the NDVI curve for each day. These predicted data were then input into various machine learning and deep learning models to predict the growth stages of winter wheat.

5.1 Cloudy Threshold Selection

As previously discussed, the percentage of "Unfounded Cloudy Days" serves as a crucial metric for assessing the efficacy and adequacy of the selected cloudy threshold. We proceed by setting the threshold range from 0.9 to 0.2 in a decrease of 0.05. In each iteration, we calculate the average percentage of "Unfounded Cloudy Days" across all fields. The outcomes of these computations are depicted in Fig. 2. The Fig. 2 illustrates a crucial observation wherein achieving a balance

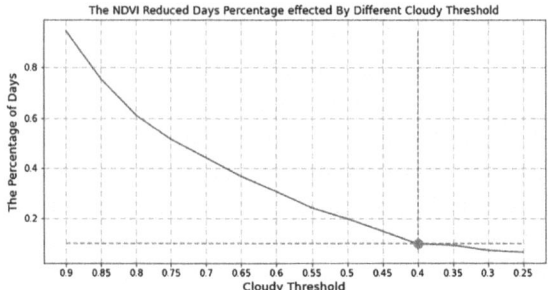

Fig. 2. The Average Possible of "Unfounded Cloudy Days" Changed with the Cloudy Threshold

between dataset size and noise reduction occurs at a cloudy threshold of 0.4. Upon reducing the threshold to 0.4, further decreases no longer significantly impact the percentage of "Unfounded Cloudy Days." Preceding this threshold, there exists an almost linear correlation between the percentage of "Unfounded Cloudy Days" and the cloudy threshold.

5.2 Model Parameters

We conducted MAPE computations for approximately 1950 fields, subsequently deriving the average MAPE to examine the variations in MAPE corresponding to increasing K values. The findings are presented in Fig. 3. Analysis of Fig. 3 reveals a discernible pattern: as the K value increases from 1 to 6, there is a consistent decrease in MAPE. However, beyond a K value of 6, there is a sudden upsurge in MAPE. Remarkably, even as the K value surpasses 10, there is a gradual yet marginal decline in MAPE, exhibiting a slower rate of decrease. Intriguingly, when K reaches 20, the MAPE still does not decrease to the same level as before when K was merely 6. Notably, our model's complexity is contingent on the product of N and K; hence, a considerable increase in K leads to a substantial rise in complexity, thereby resulting in substantial time and space

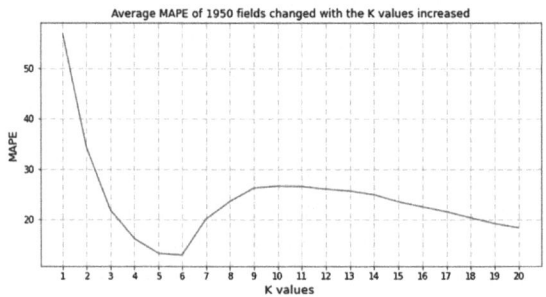

Fig. 3. Average MAPE of all fields changed by K values

waste. Consequently, for practical reasons, we opt to select K equal to 6 as the value for constructing our prediction function. Moreover, at K equal to 6, the average MAPE stands at 13.06.

Upon establishing the values for K, thereby finalizing our prediction model, we proceed to illustrate its application using the specific case of field 51258 during the year 2018, showcasing the resultant prediction outputs as presented in Fig. 4.

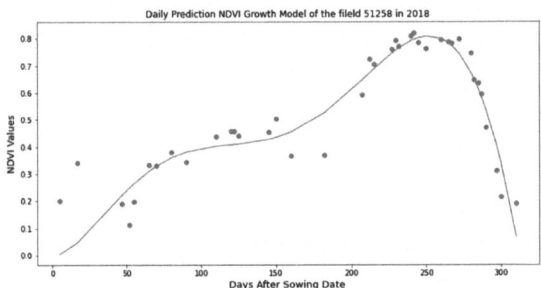

Fig. 4. Prediction of the Winter Wheat Growth

5.3 Growth Curve Prediction

This section will delineate the baseline and predictive models encompassing five distinct typologies. Their assessment will be facilitated through the Root Mean Square Error (RMSE) as a measure of evolution.

5.3.1 Baseline Model
The models introduced herein serve as the baseline against which the efficacy of our prediction model is evaluated within the pertinent literature. This baseline comprises the NDVI mean model, the utilization of ndCurveMaster to derive a singular polynomial model, and an additional Static Polynomial model implementing our prediction methodology.

– **NDVI mean value**: As previously elucidated, the NDVI mean value demonstrates effective predictability.
– **ndCurveMaster Model**: The second baseline model employs a polynomial method. In this phase, diverse models are deployed to present the findings. The initial approach entails the utilization of ndCurveMaster tools, which facilitate the division of the dataset into training and test subsets. In sequential iterations, varying numbers of coefficients are set during training (e.g., two coefficients in the first round, three in the second), and the exponents of the independent variable, DaysAfterSowing, range from 1 to 20, increasing by increments of 0.1 to identify a polynomial model minimizing RMSE.

– **Static Polynomial Model**: The second polynomial model, constructed utilizing our dataset as input, employs seven coefficients to formulate a singular polynomial model. Within this model, the entire dataset is utilized in our prediction methodology, facilitating the computation of a coefficient set tailored to fit the growth curve. This approach is similar to the ndCurveMaster model, characterized by a singular prediction line and a corresponding coefficient set for prediction. Nevertheless, a point of distinction lies in the treatment of model power: while the ndCurveMaster model is indifferent to whether the model power is integer or fractional, emphasizing solely the model order, our model predicts the Static Polynomial based on the order number yielding the smallest RMSE, with the power-constrained to integer values.

5.3.2 Dynamic Polynomial Model

As previously elucidated, we have devised a Dynamic Polynomial Model wherein, for each field, a corresponding curve and a set of coefficients are generated. Multiple sets of coefficients are at our disposal, affording the capability to ascertain the minimum and maximum values thereof. The median of the coefficients within the dynamic polynomial serves as the basis for predicting outcomes in uncharted datasets, such as those encountered in the new year, where the recorded dataset lacks complete annual data. In instances where the RMSE exhibits substantial disparities, a systematic adjustment of coefficients within a predetermined range, spanning from the minimum to the maximum values, is undertaken. This process involves iteratively modifying the coefficients from the median values by increments and decrements until the extremes are reached to identify the configuration that yields the minimal RMSE. It is noteworthy that recalibration of the model is only necessitated when a continuous escalation of RMSE is observed over three consecutive data points, obviating the need for recurrent computations with each newly recorded point. In all methodologies, the predicted NDVI values are computed, and the resulting RMSE is compared with the actual values.

Utilizing NDVI mean yields optimal outcomes, as indicated by a minimal RMSE of 0.1271. Comparable results with the ndCurveMaster or the generation of Static Polynomial, albeit their performance is inferior to the NDVI mean. The RMSE values for ndCurveMaster and Static Polynomial are 0.1675 and 0.1690, respectively, with ndCurveMaster demonstrating relatively superior performance. The dynamic polynomial model exhibits superior performance, achieving an RMSE of 0.055, significantly surpassing the baseline model. In instances where the entire annual dataset is unavailable, such as in predicting new year datasets, the adoption of median coefficient values for initializing the polynomial model is advocated. We also use this method to compute the RMSE; it can get 0.1479, which is not as good as the NDVI mean but is better than the other two polynomial baseline models. We are graphing the baseline models and prediction model depicted in Fig. 5; the yellow point represents our NDVI dataset. It is evident that the NDVI mean model most appropriately aligns with the growth trajectory of the yellow point. The green line corresponds to ndCurveMaster, while the blue line represents a Static Polynomial method. Although they converge after the apex, they exhibit considerable disparities in the initial phase. The predicted values of the Static Polynomial method are notably lower, while ndCurveMaster initially predicts excessively high values. The red line employs dynamic coefficient values of the medium coefficient to construct the polynomial model, exhibiting the closest resemblance to the NDVI mean line. This model is particularly adept at illustrating the growth patterns of winter wheat, capturing a slow increase during the winter period, spanning around 100 days, and a more rapid growth phase in the spring, commencing after 150 days. As previously delineated, the employment of the median coefficient within dynamic polynomial models as an initial predictive model for the NDVI growth curve has been expounded upon. Subsequently, in instances where the predicted data deviates significantly from the actual dataset, coefficient adjustments are requisite to accommodate the revised curve. However, extreme outliers may perturb the model's integrity. Consequently, to mitigate such anomalies, the upper 25% and lower 75% datasets were employed within our predictive methodology, augmented by ndCurveMaster, to delineate upper and lower bounds.

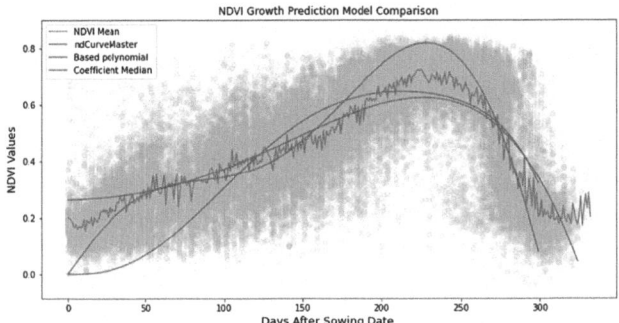

Fig. 5. NDVI Prediction Model Comparison

5.4 Growth Stages Prediction

The growth stages of winter wheat are divided into six distinct stages: sowing, emergence, stem elongation, flowering, grain filling, and ripening. According to the Wheat Growth Guide published by the Agriculture and Horticulture Development Board (AHDB)[3], these stages are defined based on Growing Degree Days (GDD). The computation starts from the sowing date, and when $GDD > 0$, it is added to the cumulative sum of degree days.

When the cumulative degree days reach 150°, leaf emergence on the main shoot begins. At 1200°, stem elongation starts, and at 2100°, the crop enters

---

[3] Agriculture and Horticulture Development Board, https://ahdb.org.uk/knowledge-library/wheat-growth-guide.

the flowering stage. Grain filling begins at 2150°, and the final ripening stage is reached at 2850°. The temperature information we collected by AdCon and ClearAg platform which have detail introduced in [22].

After segmenting the stages, our dataset consists of two input dimensions: NDVI and Slope, with the output layer representing the stage. We then divided the entire dataset into training and testing sets, randomly selecting 80% of the data as the training set and 20% as the testing set. The training set was used to train the models, while the testing set was employed to evaluate model performance. The machine learning classification supports six different target classifications, with accuracy straightforwardly computed. The DNN model output is a percentage value ranging from 0 to 1. For the DNN model, we created six models corresponding to each stage, as introduced in Sect. 4. Since the six datasets vary in size and accuracy, we tested each model and used dataset size weighting to compute the overall DNN accuracy.

The best-performing machine-learning model is the Random Forest, achieving an accuracy of 0.892. Both the KNN and Decision Tree models also exhibit strong performance, with accuracies exceeding 0.85. However, the Naive Bayes classifier is unsuitable for our dataset, as it only achieved an accuracy of 0.317.

The Deep Neural Network (DNN) model demonstrated superior performance compared to the machine learning models, with an average accuracy of 0.903. It is noteworthy that the performance of the DNN model varies across different stages. The DNN model performs exceptionally well during the sowing, emergence, and stem elongation stages, with accuracies exceeding 90%. However, its performance during the flowering stage is relatively lower, with an accuracy of 74.8%. This decline is attributed to the shorter duration of the flowering stage compared to other stages, resulting in a smaller dataset and less distinct features. The DNN model predicts the Grain Filling and Ripening stages even not over 90%, but also not very bad, with accuracy of 89.5% and 87.9%. The model achieves its best performance during the stem elongation stage, which can get 91.48%, where the growth features are most distinct.

6 Conclusion

This study encompasses analyses of both *Growth Curve Prediction* and *Growth Stages Prediction* models. Within the *Growth Curve Prediction* segment, five distinct models were evaluated, broadly categorized into baseline and prediction models. The baseline models included the NDVI mean model, the ndCurveMaster model, and the Static Polynomial model. The NDVI mean model emerged as the most effective among the baseline models, registering an RMSE of 0.1271. Our dynamic coefficient polynomial model demonstrated significantly enhanced performance, achieving an RMSE of 0.055, markedly superior to that of the baseline models.

In the *Growth Stages Prediction* segment, we evaluated five machine learning models and a DNN model. The models analyzed NDVI data and its slope from the *Growth Curve Prediction* models. Among them, the Random Forest model exhibited the highest accuracy of 89.2%. The DNN model excelled, particularly in stem elongation stage prediction, with an accuracy of 91.48%, and maintained a high average accuracy of 90.3% across all growth stages.

Acknowledgement. CONSUS is funded under the SFI Strategic Partnerships Programme (16/SPP/3296) and is co-funded by Origin Enterprises Plc.

References

1. Wang, L., Wang, J., Liu, Z., Zhu, J., Qin, F.: Evaluation of a deep-learning model for multispectral remote sensing of land use and crop classification. Crop J. **10**(5), 1435–1451 (2022)
2. Belgiu, M., Csillik, O.: Sentinel-2 cropland mapping using pixel-based and object-based time-weighted dynamic time warping analysis. Remote Sens. Environ. **204**, 509–523 (2018)
3. Li, Y., Dhelim, S., Kechadi, M.T.: Predicting winter wheat emergence and stem elongation time using CNN. In: Proceedings of the 2024 16th International Conference on Machine Learning and Computing, pp. 153–160 (2024)
4. Lin, C., Liu, Q.-S., Huang, C., Liu, G.-H.: Monitoring of winter wheat distribution and phenological phases based on modis time-series: a case study in the yellow river delta, china. J. Integr. Agric. **15**(10), 2403–2416 (2016)
5. Qu, C., Li, P., Zhang, C.: A spectral index for winter wheat mapping using multitemporal landsat NDVI data of key growth stages. ISPRS J. Photogramm. Remote. Sens. **175**, 431–447 (2021)
6. Singh, B.M., Komal, C., Victorovich, K.A.: Crop growth monitoring through sentinel and landsat data based NDVI time-series. Comput. Opt. **44**(3), 409–419 (2020)
7. Li, C., et al.: Using NDVI percentiles to monitor real-time crop growth. Comput. Electron. Agric. **162**, 357–363 (2019)
8. Liu, K.-L., Li, Y.-Z., Hu, H.-W.: Predicting ratoon rice growth rhythm based on NDVI at key growth stages of main rice. Chil. J. Agric. Res. **75**(4), 410–417 (2015)
9. Herbei, M.V., Florin, S.: Use landsat image to evaluate vegetation stage in sunflower crops. AgroLife Sci. J. **4**(1) (2015)
10. Stăncescu, A., Sala, F.: Comparative analysis of two satellite systems in services for agriculture. case study: the usamvbt teaching and experimental resort. Res. J. Agric. Sci. **51**(2) (2019)
11. Domínguez, J.A., Kumhálová, J., Novák, P.: Winter oilseed rape and winter wheat growth prediction using remote sensing methods. Plant Soil Environ. **61**(9), 410–416 (2015)
12. Shammi, S.A., Meng, Q.: Use time series NDVI and EVI to develop dynamic crop growth metrics for yield modeling. Ecol. Ind. **121**, 107124 (2021)
13. Seo, B., Lee, J., Lee, K.-D., Hong, S., Kang, S.: Improving remotely-sensed crop monitoring by NDVI-based crop phenology estimators for corn and soybeans in Iowa and Illinois, USA. Field Crop. Res. **238**, 113–128 (2019)
14. Berger, A., Ettlin, G., Quincke, C., Rodríguez-Bocca, P.: Predicting the normalized difference vegetation index (NDVI) by training a crop growth model with historical data. Comput. Electron. Agric. **161**, 305–311 (2019)
15. Jayawardhana, W., Chathurange, V.: Extraction of agricultural phenological parameters of Sri Lanka using MODIS, NDVI time series data. Procedia Food Sci. **6**, 235–241 (2016)

16. Zhang, L., Zhang, Z., Luo, Y., Cao, J., Xie, R., Li, S.: Integrating satellite-derived climatic and vegetation indices to predict smallholder maize yield using deep learning. Agric. For. Meteorol. **311**, 108666 (2021)
17. Son, N.-T., et al.: Machine learning approaches for rice crop yield predictions using time-series satellite data in Taiwan. Int. J. Remote Sens. **41**(20), 7868–7888 (2020)
18. Stas, M., Van Orshoven, J., Dong, Q., Heremans, S., Zhang, B.: A comparison of machine learning algorithms for regional wheat yield prediction using NDVI time series of spot-VGT. In: Fifth International Conference on Agro-Geoinformatics (Agro-Geoinformatics), pp. 1–5. IEEE (2016)
19. Yue, Y., et al.: Prediction of maize growth stages based on deep learning. Comput. Electron. Agric. **172**, 105351 (2020)
20. Teodoro, P.E., et al.: Predicting days to maturity, plant height, and grain yield in soybean: a machine and deep learning approach using multispectral data. Remote Sens. **13**(22), 4632 (2021)
21. Jeong, S., Ko, J., Shin, T., Yeom, J.-M.: Incorporation of machine learning and deep neural network approaches into a remote sensing-integrated crop model for the simulation of rice growth. Sci. Rep. **12**(1), 9030 (2022)
22. Li, Y., Zainab, S.S.E., Dhelim, S., Kechadi, M.T.: Weather data and agro-climate indices–comparative study. In: Proceedings of the 2024 16th International Conference on Machine Learning and Computing, pp. 1–7 (2024)

Cross-Lingual Style Transfer TTS for High-Quality Machine Dubbing

Stoyan Mihov[✉]

Institute of Information and Communication Technologies, BAS, Sofia, Bulgaria
stoyan@lml.bas.bg
http://lml.bas.bg/~stoyan

Abstract. Style transfer is essential for high-quality machine dubbing. While numerous approaches for style transfer in speech synthesis have been developed, cross-lingual style transfer remains a significant challenge. In this paper we introduce a novel speech synthesis method which realizes style learning across different languages and speakers. Our approach features a transformer-based architecture with a speech prompted text encoder, a duration predictor and a flow matching generative decoder. The text encoder is conditioned on the noisy source language speech, which is entered as speech prompt for style adaptation. The flow matching generative decoder produces high quality speech conditioned on the text encoder output. Empirical evaluations demonstrate that our TTS system generates speech that is objectively closer to the recordings of professional voice talents compared to a strong baseline model. Audio samples are available on our demo page (http://lml.bas.bg/~stoyan/dubbing).

Keywords: Machine dubbing · Style transfer · Text-to-speech synthesis

1 Introduction

In recent years, the rapid advancement of text-to-speech (TTS) technologies has significantly transformed various applications, ranging from virtual assistants and audiobooks to automated customer service and accessibility tools. Among these advancements, one particularly compelling area of research is machine dubbing, which involves the automatic vocalization of audiovisual content into different languages. Traditional methods for TTS and machine dubbing often struggle with preserving the speaker's speaking rate and prosody across languages, resulting in unconvincing outputs.

Research in style transfer for text-to-speech synthesis has evolved substantially, with early efforts focusing primarily on improving the naturalness and intelligibility of synthesized speech. Recent research has increasingly emphasized style transfer, aiming to modify and adapt the prosody, emotion, and speaker identity in synthesized speech. However, achieving high-quality cross-lingual style transfer remains a complex challenge due to the intricate interplay between linguistic content and stylistic expression.

Brannon et al. presented a significant study contributing to the field of style transfer for machine dubbing in [1]. The researchers examined various aspects of human dubbing, including the intricacies of matching lip movements, maintaining consistent speaker identity, and preserving emotional tone. Their findings highlight the challenges faced in human dubbing, underscoring the necessity for advanced techniques that can replicate the nuanced performance of human dubbers. The study emphasizes that effective automatic dubbing systems must integrate sophisticated mechanisms for style transfer and precise timing and synchronization to achieve human-like quality and expressiveness.

One of the most promising approaches in this domain involves the use of speech prompts for cross-lingual style transfer. This method leverages a reference speech sample to guide the TTS system in generating speech that mirrors the style, intonation, and emotion of the original audio, even in a different language. In this article, we present a novel approach to text-to-speech synthesis that incorporates cross-lingual style transfer via a speech prompt technique. We explore the underlying methodology and its comparative advantages over a strong baseline approach.

2 Related Works

The field of style transfer in text-to-speech (TTS) synthesis has witnessed significant advancements over the past decade, driven by the development of deep learning techniques. Early works in TTS primarily focused on improving the naturalness and intelligibility of synthesized speech. Notable examples include the Tacotron [23] and FastSpeech [18,19] series. In the last 2–3 years research has been focused towards handling the one-to-many mapping in TTS by applying more sophisticated generative methods like Invertible Flows – GlowTTS [7], Variational Autoencoders – VITS [8], Denoising Diffusion Probabilistic Models – GradTTS [16], and others. In the last year the Flow Matching for Generative Modelling approach – P-Flow [9], VoiceBox [12], MATCHA-TTS [15] set new benchmarks for speech quality and expressiveness.

Building upon these foundations, researchers are exploring methods to transfer stylistic attributes such as prosody, emotion, and speaker identity. Prosody transfer techniques involve transferring the rhythm and intonation patterns from one speaker's speech to another. Skerry-Ryan et al. introduced in [20] a method that captures prosody at the utterance level, enabling more expressive and natural speech synthesis. Wang et al. [24] proposed Global Style Tokens (GSTs) that allow the model to learn and control different speaking styles in a unified framework. AdaSpeech, introduced by Chen et al. in [2], further advanced the field by offering a highly adaptive TTS model capable of fine-grained control over the synthesized speech style.

Cross-lingual style transfer for machine dubbing introduces additional complexities due to the need to maintain stylistic consistency while translating linguistic content across languages. The paper [17] by Rattcliffe et al. presents an approach to cross-lingual style transfer in text-to-speech synthesis using a Conditional Prior Variational Autoencoder combined with a style loss mechanism. Effendi et al. in [3] focus on the crucial aspect of duration modelling in neural TTS systems for automatic dubbing. Recent advancements have focused on leveraging speech prompts for cross-lingual style transfer. The use of speech prompts involves using an audio sample to guide the synthesis process. This method has shown significant potential in retaining the speaker's style and emotional tone across languages. Chen et al. introduced in [22] a speech prompt-based TTS system for transferring the style from the prompt to the synthesized speech in a different language. In their approach, however, the speech prompt is encoded into a fix-sized vector, what limits its effectiveness for phrase-level speaking rate transfer. Swiatkowski et al. extend this method in [21] by enabling modelling and cross-lingual transfer of prosody at phrase level. However, the audio segmentation and the alignment of phrase-level audio reference to target text phonemes is a challenging task.

These related works highlight the ongoing efforts and innovations in style transfer for TTS and cross-lingual applications. The proposed method aims to address the remaining challenges in achieving high-quality, expressive, and natural-sounding machine-dubbed speech across different languages.

3 Method Description

In this work the aim is to propose a TTS system with the capabilities of learning the speech style from the original speech in the source language. For this purpose we apply a prompting approach which utilizes the noisy reference audio in source language spoken by a source speaker to generate speech in the target language with the voice characteristics of a target speaker. First, the reference audio prompt, the target text and the target speaker ID are processed by a text encoder. Afterwards, the upsampled output of the text encoder is conditioning a flow matching decoder to generate the target speech. In general the proposed model architecture is similar to the P-Flow TTS [9].

3.1 Overview of the Flow Matching Method

In this subsection we briefly present the flow matching method closely following the description from Lipman et al. [13]. The flow matching method aims to model a path in time between a simple known prior probability density $p_0(x)$ and the observed density of the data $p_1(x)$. We assume that the prior density is standard Gaussian i.e. $p_0(x) = \mathcal{N}(x \,|\, 0, I)$. We define the flow as $\phi : [0, 1] \times \mathbb{R}^d \to \mathbb{R}^d$, such that $\phi_0(x) = x$. The probability density p_t for $t \in [0, 1]$ is defined by variable substitution:

$$p_t(x) = p_0(\phi_t^{-1}(x)) \det \left[\frac{\partial \phi_t^{-1}}{\partial x}(x) \right].$$

The flow ϕ_t is constructed from the vector field v_t via the Ordinary Differential Equation (ODE):

$$\frac{d}{dt}\phi_t(x) = v_t(\phi_t(x))$$
$$\phi_0(x) = x$$

We model the vector field v_t with a neural network $v_t(x;\theta)$, where $\theta \in \mathbb{R}^p$ are the network parameters to be learned. The objective is to find the parameters θ such that the vector field v_t generates a probability path p_t, which flows from p_0 to p_1. If we know a vector field $u_t(x)$ which generates the probability density path $p_t(x)$, then we can define the Flow Matching (FM) objective as

$$\mathcal{L}_{FM}(\theta) = \mathbb{E}_{t, p_t(x)} \|v_t(x;\theta) - u_t(x)\|^2,$$

where $t \sim \mathcal{U}[0,1]$ and $x \sim p_t(x)$. But in general p_t and u_t are not known.

Let us denote the unknown probability distribution of the data as $q(x_1)$. Let us define the conditional probability paths as

$$p_t(x|x_1) = \mathcal{N}(x \mid \mu_t(x_1), \sigma_t(x_1)^2 I),$$

where $\mu : [0,1] \times \mathbb{R}^d \to \mathbb{R}^d, \mu_0(x_1) = 0$ and $\sigma : [0,1] \times \mathbb{R}^d \to \mathbb{R}^+, \sigma_0(x_1) = 1$. In that case we have $p_0(x|x_1) = \mathcal{N}(x \mid 0, I)$. Further we assume that $\mu_1(x_1) = x_1$ and $\sigma_1(x_1) = \sigma_{min}$ for some sufficiently small $\sigma_{min} > 0$. In that case $p_1(x|x_1)$ is a concentrated Gaussian distribution with centre x_1. We define

$$p_t(x) = \int p_t(x|x_1) q(x_1) dx_1.$$

For the path p_t holds $p_0(x) = \mathcal{N}(x \mid 0; I)$ and $p_1(x) \approx q(x)$. Let us assume that the conditional vector field $u_t(\cdot|x_1) : \mathbb{R}^d \to \mathbb{R}^d$ generates the conditional probability distribution $p_t(\cdot|x_1)$. Then it is shown that the vector field

$$u_t(x) = \int u_t(x|x_1) \frac{p_t(x|x_1) q(x_1)}{p_t(x)} dx_1$$

generates the probability path p_t.

We define the Conditional Flow Matching (CFM) objective as

$$\mathcal{L}_{CFM}(\theta) = \mathbb{E}_{t, q(x_1), p_t(x|x_1)} \|v_t(x;\theta) - u_t(x|x_1)\|^2,$$

where $t \sim \mathcal{U}[0,1], x_1 \sim q(x_1)$ $x \sim p_t(x|x_1)$. It can be shown that $\mathcal{L}_{FM}$ and $\mathcal{L}_{CFM}$ are equal up to a constant independent of θ. Hence, $\nabla_\theta \mathcal{L}_{FM}(\theta) = \nabla_\theta \mathcal{L}_{CFM}(\theta)$.

Optimal Transport (OT) Conditional Vector Fields. A natural and effective choice for the conditional probability paths is to change the mean and deviation linearly in time:

$$\mu_t(x_1) = tx_1, \text{ and } \sigma_t(x_1) = 1 - (1 - \sigma_{min})t.$$

In that case the closed form of the OT conditional vector field is

$$u_t(x|x_1) = \frac{x_1 - (1 - \sigma_{min})x}{1 - (1 - \sigma_{min})t},$$

and the flow corresponding to the vector field $u_t(x|x_1)$ is

$$\psi_t(x) = (1 - (1 - \sigma_{min})t)x + tx_1.$$

In the case of OT conditional vector field the CFM objective takes the form

$$\mathcal{L}_{CFM}(\theta) = \mathbb{E}_{t,q(x_1),p_0(x_0)} \|v_t(\psi_t(x_0);\theta) - (x_1 - (1 - \sigma_{min})x_0)\|^2.$$

In our method we apply the CFM objective for the OT conditional vector field given above.

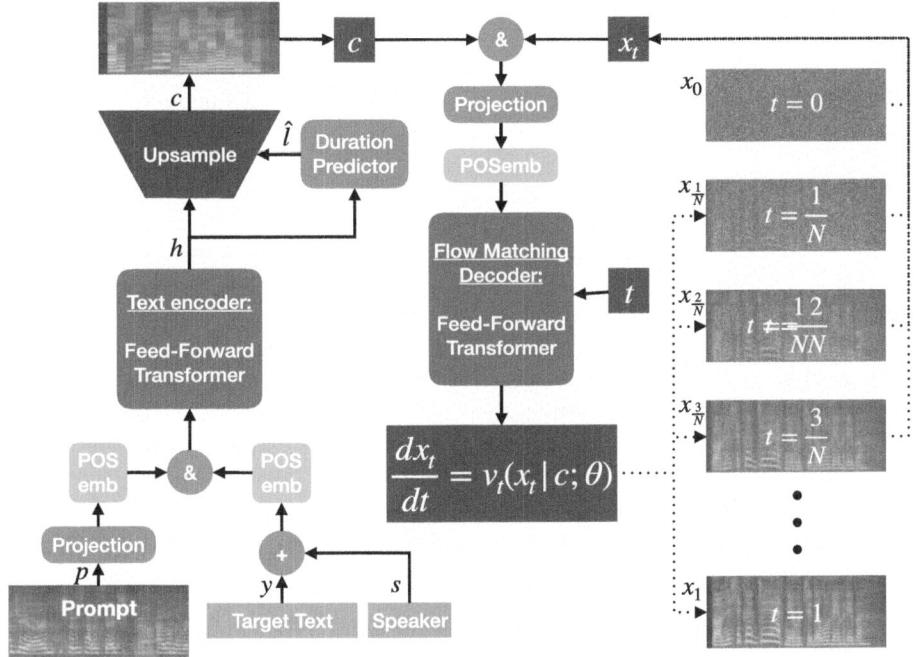

Fig. 1. The overall diagram of the proposed approach at synthesis time.

3.2 Model Architecture

The diagram on Fig. 1 presents the architecture of our model.

Let s be the target speaker identifier, $p = (p^1, p^2, \ldots, p^K)$ be the reference speech prompt of K frames, $y = (y^1, y^2, \ldots, y^M)$ be a text sequence of M phones,

$l = (l^1, l^2, \ldots, l^M)$ be the per-phone duration where l^j denotes how many audio frames y^j correspond to and $x = (x^1, x^2, \ldots, x^N)$ be the target audio of N frames s.t. $\sum_{j=1}^M l^j = N$. In practice $K \approx N$ because the source and target audio have similar durations.

First, the embedded target text y and the embedded speaker identifier s are added together. The reference speech prompt p is projected to the encoder dimension. Afterwards, positional encodings are added to the speech prompt and the target text. The positional encoding is defined as the sum of absolute positional encoding and a learnable fixed-size embedding so that the encoder can differentiate the speech prompt and text through the learnable embeddings. The speech prompt and the target text are concatenated in the time dimension and used as inputs to the text encoder. The goal of the text encoder is to generate a text representation c, using the reference speech prompt for style transfer and the target speaker embedding for speaker information. The text encoder employs a non-autoregressive transformer architecture that can attend to the speech prompt at arbitrary text positions. The output of the encoder is $h = f_{\text{enc}}(y, s, p; \theta) = (h^1, h^2, \ldots, h^M)$.

We are using an explicit duration learning module for learning the phone durations from forced alignments as in [18]. After the text encoder we apply a duration predictor with output $\hat{l} = f_{\text{dur}}(h; \theta)$. We define the duration loss as

$$\mathcal{L}_{\text{dur}}(\theta) = \mathbb{E}_{q(l,h)} \|l - f_{\text{dur}}(h; \theta)\|^2.$$

We obtain the text representation c by upsampling the output of the text encoder h, i.e. we define $c = \text{rep}(h, l) = (c^1, c^2, \ldots, c^N)$ to be the frame-level transcript, which repeats each h^j by l^j times such that c^i denotes the text representation of the audio frame x^i. The text representation c is then used for conditioning the flow matching decoder. Let $x_t = (x_t^1, x_t^2, \ldots, x_t^N)$ be the sample at flow step t. We sample $x_0 \sim \mathcal{N}(x \,|\, 0, I)$. At training time we have $x_t = \psi_t(x_0) = (1 - (1 - \sigma_{min})t)x_0 + tx$. We concatenate x_t with the condition c in the data dimension and project the result to the decoder dimension to be used as input to the flow matching decoder. The decoder is additionally conditioned on the time step t. In our setup the output of the decoder is $v_t(x_t, c; \theta)$ and the CFM objective is

$$\mathcal{L}_{CFM}(\theta) = \mathbb{E}_{t, q(x,c), p_0(x_0)} \|v_t(x_t, c; \theta) - (x - (1 - \sigma_{min})x_0)\|^2.$$

Our model has in total 26.3 M parameters. Below we give the details for the main modules.

Speech-Prompted Text Encoder. Our text encoder is a 6-layer Feed Forward Transformer (FFT) [18] of 384 hidden dimensions with 2 attention heads. The linear steps of the transformer uses depth-wise separable convolutions [6] for memory and computation efficiency as in [14]. The size of the convolution kernels is 9. The text encoder has 10.7 M parameters.

Flow Matching Decoder. The proposed flow matching decoder utilizes a very similar architecture consisting of a 6-layer Feed Forward Transformer [18] of 384 hidden dimensions with 2 attention heads and depth-wise separable convolutions with kernel size 9 for the linear steps. The FFT is amended with Conditional Layer Normalization [2] for conditioning on positional embedding of the scaled time step t. The output of the transformer is projected to the frame dimension by applying two linear transformations with intermediate hidden dimension of 1536 and a ReLU activation in between. The number of parameters of the decoder is 15 M.

Duration Predictor. The duration predictor is of 256 hidden dimensions and consists of 3 layers of convolutions with kernel size 3. The convolutions are followed by ReLU, layer normalization and dropout with $p = 0.1$ as in [18]. The duration predictor has 0.5 M parameters.

3.3 Model Training

The model is trained by minimizing the total loss

$$\mathcal{L}(\theta) = \mathcal{L}_{\text{dur}}(\theta) + \mathcal{L}_{CFM}(\theta) = \\ = \mathbb{E}_{q(l,h)} \|l - f_{\text{dur}}(h;\theta)\|^2 + \mathbb{E}_{t,q(x,c),p_0(x_0)} \|v_t(x_t,c;\theta) - (x - (1-\sigma_{min})x_0)\|^2,$$

where the data point (x, y, s, p, l) is sampled from the training dataset, $h = f_{\text{enc}}(y, s, p; \theta)$, $c = \text{rep}(h, l)$, $t \sim \mathcal{U}[0,1]$, $x_0 \sim \mathcal{N}(x \mid 0, I)$ and $x_t = \psi_t(x_0)$.

3.4 Inference

The inference procedure takes as input the target text y, reference prompt p and speaker identifier s. First we obtain the text encoder output $h = f_{\text{enc}}(y, s, p; \theta)$ and the duration prediction $\hat{l} = f_{\text{dur}}(h; \theta)$. Using those we get the condition $c = \text{rep}(h, \hat{l})$. Then we sample $x_0 \sim \mathcal{N}(x \mid 0, I)$. Afterwards we find $x_1 = \phi_1(x_0)$ by solving the ODE

$$\frac{d}{dt}\phi_t(x) = v_t(\phi_t(x), c; \theta)$$
$$\phi_0(x_0) = x_0.$$

For solving the ODE we use the first-order Euler's method with e.g. $N = 20$ steps. Euler's method with N steps is performed starting from $t = 0$ and applying the following recurrent formula:

$$x_{t+\frac{1}{N}} = x_t + \frac{1}{N} v_t(x_t, c; \theta).$$

Classifier-Free Guidance. As in [9] we enhance further the pronunciation clarity by applying techniques from a classifier-free guidance method [5]. As in [9] we compute $\bar{c}$ by averaging c along the time axis to obtain a fixed-size vector and then duplicating it along the time axis. Then the guidance-amplified Euler formula is

$$x_{t+\frac{1}{N}} = x_t + \frac{1}{N}\left((1+\gamma)v_t(x_t, c; \theta) - \gamma v_t(x_t, \bar{c}; \theta)\right),$$

where $\gamma > 0$ is the guidance scale. We use $\gamma = 1$.

4 Experiments

In order to evaluate the proposed TTS system we prepared a relevant dataset. We also developed a strong baseline model, which does not make use of a reference speech prompt. We trained both models on the dataset and produced the target speech with both models for evaluation. Finally we compare the result using objective metrics.

4.1 Dataset

For producing a relevant dataset we process a set of 614 TV episodes from various TV shows with a total duration of 440 h. For all TV episodes we have the original audio tracks in English language. The original audio tracks are noisy – they contain a mixture of the actors' speeches and the environment noises. In addition to the original audio tracks we have dubbed Bulgarian audio tracks which contain the clean speech of the voice talents and the transcripts translated in Bulgarian. The TV shows have been dubbed by 40 individual speakers. The cumulative speech duration of each speaker varies between 4 min and 50 h.

For building the speech corpus we use the methodology presented in [4]. First the texts in the transcripts are normalized. Then the Bulgarian audio tracks are force-aligned with the transcripts using automatic speech recognition (ASR). Further the audio and texts are cleaned and segmented into parallel utterances.

We next estimate the phoneme level durations through forced alignment using a speech recognition model. Finally, for each utterance the reference English speech prompt is extracted from the original audio track by cutting the corresponding time interval.

In the training dataset compilation two TV episodes are left out to be used later for testing. We also limit the cumulative speech of each speaker to 12 h. The resulting training dataset consists of 227500 utterances with total duration of 150 h.

4.2 Baseline Model

For the baseline model, we employ the same flow matching architecture for speech generation as used in our proposed model. This baseline model is strong in the sense that it utilizes the state-of-the-art flow matching generative architecture, has similar number of parameters and is trained in the same way as the proposed method. The key difference is the absence of the reference speech prompt, along with its associated projection and positional embedding. The total number of parameters in the baseline model is 26.27 M. This baseline model allows us to objectively assess the impact of incorporating the reference speech prompt.

4.3 Training and Inference Settings

The proposed and the baseline models are trained for 1.5 M iterations, using a batch size of 128. We utilize the Adam optimizer [10] with $\beta_1 = 0.9$, $\beta_2 = 0.999$ and $\epsilon = 10^{-6}$. The learning rate starts at 0.0002 and decays exponentially by 0.9999977 at each iteration. During inference, we generate Mel-spectrograms using 20 Euler steps in the flow matching decoder with a guidance scale of 1.0. We generate the waveforms from the Mel-spectrograms using the Hifi-GAN vocoder [11], trained on the same dataset. Our audio representation is 24 kHz audio represented with an 100-bin Mel-spectrogram, with window size 1024 and hop length 256.

4.4 Evaluation

For evaluation we use the utterances from one TV episode, which is not included in the training set. The utterances are filtered to satisfy two constraints: (i) to have length of more than 15 symbols; and (ii) to have exact matches of the audio ASR with the transcripts. In total 102 utterances have been retrieved for testing. The average duration of the test utterances is 2.33 s. For each of the utterances we prepare the original English audio prompt and the Bulgarian human dubbed utterance. The synthesized speeches are generated by specifying the ID of the speaker who has performed the actual human dubbing. This setup allows us to make a direct comparison of the human dubbed speech with the synthesized speech for the same speaker conditioned on the same original reference speech.

We present samples for subjective evaluation on our demo page: http://lml.bas.bg/~stoyan/dubbing.

Table 1. Mean utterance duration differences with 95% CI

	Human dubbed	Proposed (w. prompt)	Baseline (w/o prompt)
Original audio	319 ms ± 71	**104 ms ± 25**	522 ms ± 71
Human dubbed	—	**334 ms ± 76**	466 ms ± 104

4.5 Objective Metrics

In order to assess the isochrony we first compare the duration differences between the synthesized and the recorded human speeches.

We show in Table 1 the mean of the utterance duration differences with respect to the corresponding human speech. For each mean value we give the 95% Confidence Interval (CI). The comparison shows that the duration difference of the dubbed with the original speech is 319 ms in average. The duration difference of the utterances generated by the proposed model against the original audio is 104 ms. In that respect, the baseline method is around 5 times less accurate. Compared against the human dubbed audio, our method is around 28% more accurate than the baseline.

Our next experiment aims to evaluate the phone duration differences. In order to have a fair experiment we first ensure that the durations of the synthesized utterances are equal to the durations of the corresponding human dubbed utterances. This is done by scaling the predicted phone durations by the corresponding factor. In Table 2 we compare the phone duration differences after scaling. Here the proposed method is by 10% more precise than the baseline.

Our last experiment aims to objectively compare the prosody of the generated speech. As the objective evaluation metrics we use Root Mean Square Error (RMSE) of $\log F_0$ and Mel-Cepstrum Distortion (MCD). We use the corresponding functions from the ESPnet toolkit [25] to perform the calculations. Again, in order to have a fair experiment and to align the synthesized audio with the human dubbed speech we control the phone durations. For this experiment we force the phone durations to exactly match the phone durations of the human dubbed speech. The results of the $\log F_0$ and MCD comparison after forcing the alignment are presented in Table 3. The table shows that our new method delivers 5.9% lower $\log F_0$ RMSE and 8.9% lower MCD against the human dubbed speech in comparison to the baseline method.

Table 2. Mean phone duration differences with 95% CI

	Proposed (w. prompt)	Baseline (w/o prompt)
Human dubbed	**22.9 ms** ± 1.9	25.5 ms ± 2.3

Table 3. $\log F_0$ RMSE and MCD with 95% CI

	Proposed (w. prompt)	Baseline (w/o prompt)
$\log F_0$ RMSE	**0.301** ± 0.022	0.320 ± 0.023
MCD	**8.14** ± 0.189	8.94 ± 0.239

5 Discussion

We have introduced a novel TTS method for high-quality machine dubbing, leveraging reference utterances in the source language as prompts for effective cross-lingual style adaptation. Our approach significantly improves isochrony, with predicted durations that are notably more precise than those produced by a strong baseline. Furthermore, our method aligns more closely with the original speech durations compared to human-dubbed speeches. Objective prosody evaluations demonstrate that our system outperforms a strong baseline model, a finding confirmed by informal subjective tests.

Unlike other approaches [21, 22], our architecture is using directly raw, noisy reference audio without requiring explicit sub-utterance alignment between the reference audio and target text. We speculate that the powerful transformer architecture of the text encoder effectively learns to extract speech features from the noisy prompt. Additionally, the attention mechanism within the transformer blocks learns the alignment between the reference audio and the target text.

In addition to improved timing and prosody, our method generates speech with better maintenance of the stylistic nuances and emotional tone of the original utterances. Future work will explore further optimization of the model for real-world applications and integration of additional linguistic features to enhance the naturalness and expressiveness of the synthesized speech.

Acknowledgments. We acknowledge the provided access to the e-infrastructure of the Centre for Advanced Computing and Data Processing, with the financial support by the Grant No BG05M2OP001-1.001-0003, financed by the Science and Education for Smart Growth Operational Program (2014-2020) and co-financed by the European Union through the European structural and Investment funds.

The reported work has been partially supported by CLaDA-BG, *the Bulgarian National Interdisciplinary Research e-Infrastructure for Resources and Technologies in favor of the Bulgarian Language and Cultural Heritage, part of the EU infrastructures CLARIN and DARIAH*, funded by the Ministry of Education and Science of Bulgaria (support for the Bulgarian National Roadmap for Research Infrastructure).

References

1. Brannon, W., Virkar, Y., Thompson, B.: Dubbing in practice: a large scale study of human localization with insights for automatic dubbing. Trans. Assoc. Comput. Linguist. **11**, 419–435 (2023)
2. Chen, M., et al.: Adaspeech: adaptive text to speech for custom voice. In: International Conference on Learning Representations (2021)
3. Effendi, J., Virkar, Y., Barra-Chicote, R., Federico, M.: Duration modeling of neural TTS for automatic dubbing. In: ICASSP 2022 (2022)
4. Geneva, D., Shopov, G., Mihov, S.: Building an ASR corpus based on Bulgarian parliament speeches. In: Martín-Vide, C., Purver, M., Pollak, S. (eds.) SLSP 2019. LNCS (LNAI), vol. 11816, pp. 188–197. Springer, Cham (2019). https://doi.org/10.1007/978-3-030-31372-2_16

5. Ho, J., Salimans, T.: Classifier-free diffusion guidance. In: NeurIPS 2021 Workshop on Deep Generative Models and Downstream Applications (2021)
6. Kaiser, L., Gomez, A.N., Chollet, F.: Depthwise separable convolutions for neural machine translation. In: International Conference on Learning Representations (2018)
7. Kim, J., Kim, S., Kong, J., Yoon, S.: Glow-TTS: a generative flow for text-to-speech via monotonic alignment search. In: Proceedings of the 34th International Conference on Neural Information Processing Systems. NIPS 2020. Curran Associates Inc. (2020)
8. Kim, J., Kong, J., Son, J.: Conditional variational autoencoder with adversarial learning for end-to-end text-to-speech. In: Meila, M., Zhang, T. (eds.) Proceedings of the 38th International Conference on Machine Learning. Proceedings of Machine Learning Research, vol. 139, pp. 5530–5540. PMLR (2021)
9. Kim, S., et al.: P-flow: a fast and data-efficient zero-shot TTS through speech prompting. In: Thirty-Seventh Conference on Neural Information Processing Systems (2023)
10. Kingma, D.P., Ba, J.: Adam: a method for stochastic optimization. In: Bengio, Y., LeCun, Y. (eds.) 3rd International Conference on Learning Representations, ICLR 2015, San Diego, CA, USA, 7–9 May 2015, Conference Track Proceedings (2015)
11. Kong, J., Kim, J., Bae, J.: Hifi-gan: generative adversarial networks for efficient and high fidelity speech synthesis. In: Larochelle, H., Ranzato, M., Hadsell, R., Balcan, M., Lin, H. (eds.) Advances in Neural Information Processing Systems 33: Annual Conference on Neural Information Processing Systems 2020, NeurIPS 2020, 6–12 December 2020, virtual (2020)
12. Le, M., et al.: Voicebox: text-guided multilingual universal speech generation at scale. In: Thirty-Seventh Conference on Neural Information Processing Systems (2023)
13. Lipman, Y., Chen, R.T.Q., Ben-Hamu, H., Nickel, M., Le, M.: Flow matching for generative modeling. In: The Eleventh International Conference on Learning Representations (2023)
14. Luo, R., et al.: Lightspeech: lightweight and fast text to speech with neural architecture search. In: ICASSP 2021 - 2021 IEEE International Conference on Acoustics, Speech and Signal Processing (ICASSP), pp. 5699–5703 (2021)
15. Mehta, S., Tu, R., Beskow, J., Szèkely, E., Henter, G.E.: Matcha-TTS: a fast TTS architecture with conditional flow matching. In: ICASSP 2024 - 2024 IEEE International Conference on Acoustics, Speech and Signal Processing (ICASSP), pp. 11341–11345 (2024)
16. Popov, V., Vovk, I., Gogoryan, V., Sadekova, T., Kudinov, M.: Grad-TTS: a diffusion probabilistic model for text-to-speech. In: Meila, M., Zhang, T. (eds.) Proceedings of the 38th International Conference on Machine Learning. Proceedings of Machine Learning Research, vol. 139, pp. 8599–8608. PMLR (2021)
17. Ratcliffe, D., Wang, Y., Mansbridge, A., Karanasou, P., Moinet, A., Cotescu, M.: Cross-lingual style transfer with conditional prior VAE and style loss. In: Interspeech 2022 (2022)
18. Ren, Y., et al.: Fastspeech 2: fast and high-quality end-to-end text to speech. arXiv preprint arXiv:2006.04558 (2020)
19. Ren, Y., et al.: Fastspeech: fast, robust and controllable text to speech. In: Advances in Neural Information Processing Systems, vol. 32 (2019)
20. Skerry-Ryan, R., et al.: Towards end-to-end prosody transfer for expressive speech synthesis with tacotron. In: Dy, J., Krause, A. (eds.) Proceedings of the 35th

International Conference on Machine Learning. Proceedings of Machine Learning Research, vol. 80, pp. 4693–4702. PMLR (2018)
21. Swiatkowski, J., et al.: Expressive machine dubbing through phrase-level cross-lingual prosody transfer. In: Proceedings of INTERSPEECH 2023, pp. 5546–5550 (2023). https://doi.org/10.21437/Interspeech.2023-441
22. Swiatkowski, J., Wang, D., Babianski, M., Lumban Tobing, P., Vipperla, R., Pollet, V.: Cross-lingual prosody transfer for expressive machine dubbing. In: Proceedings of INTERSPEECH 2023, pp. 4838–4842 (2023). https://doi.org/10.21437/Interspeech.2023-437
23. Wang, Y., et al.: Tacotron: towards end-to-end speech synthesis. In: Proceedings of Interspeech 2017, pp. 4006–4010 (2017)
24. Wang, Y., et al.: Style tokens: unsupervised style modeling, control and transfer in end-to-end speech synthesis. In: Dy, J., Krause, A. (eds.) Proceedings of the 35th International Conference on Machine Learning. Proceedings of Machine Learning Research, vol. 80, pp. 5180–5189. PMLR (2018)
25. Watanabe, S., et al.: Espnet: end-to-end speech processing toolkit. In: Proceedings of INTERSPEECH 2018, pp. 2207 – 2211 (2018)

An Approach to Discovering, Tracking Over Time, and Summarizing Publicly Available Information on a Given Topic

Alexandrina Karakehayova and Maria Nisheva-Pavlova[✉]

Faculty of Mathematics and Informatics, Sofia University, St. Kliment Ohridski, 5 James Bourchier Blvd., 1164 Sofia, Bulgaria
akarakehaj@uni-sofia.bg, marian@fmi.uni-sofia.bg

Abstract. The paper presents a pilot study aimed at developing a novel approach for discovering, tracking over time, and summarizing the content of available publications on a given significant topic. A key result of this study is the proposed methodology for summarizing information from multiple articles on a specified topic by using large language models like BART and GPT with the help of fine tuning and prompt engineering. The developed approach has been applied in the creation of a software tool that collects publicly available information on a given topic from reliable sources for a specified period of time and displays in a concise form the most essential of the content of the publications found, thereby tracking the history of information on the topic.

Keywords: Information Retrieval · Multiple Texts Summarization · Natural Language Processing · Machine Learning · Large Language Model · Prompt Engineering

1 Introduction

Information is one of the most valuable tools nowadays, and obtaining high-quality information in a concise form is priceless in our intensive lives. This study focuses on the development and experimentation of a comprehensive approach to filter reliable sources from the vast data available on the Internet and provide in a condensed format up-to-date and relevant information tailored to the user's search query. As a result, a pilot software tool has been developed which offers a convenient user interface where users can input a word, phrase, or question to perform the search, select the period from which the information should be retrieved, and choose the model to be used for summarization.

The final result is presented along with links to the original articles for reference. The news is gathered daily from reputable media outlets such as CNN, BBC, and Associated Press. Using large language models (LLMs) like BART [1] and GPT [2] with the help of fine-tuning and prompt engineering techniques, the software successfully summarizes multiple news articles and shows in a concise form the most essential, in particular the changes over time of information on the topic, to save the user time and effort spent on searching and analyzing texts. By continuing to refine these models using novel approaches, our work further enhances their capabilities, making them valuable tools in the field of natural language processing and beyond.

2 Related Work

Some of the tasks of our research find a solution in various existing specialized semantic search engines, but there is no one that implements the whole set of them. However, we will discuss some of the alternatives with their advantages and disadvantages.

2.1 Gathering News from Various Sources in Real Time

Among the most popular solutions to this problem can be mentioned SmartNews[1] and Pocket[2].

According to the official website of SmartNews, the aim of the product is to provide quality information to users through the best algorithms and a simple user interface. It offers a mix of news from multiple sources, using both machine learning algorithms to find relevant articles against the search query and recommender systems to suggest relevant information to the user. However, there is no summarized form of the received information. The user is referred to the original news, and if the corresponding website requires a subscription or registration, it cannot be accessed. Also, the app is mobile only and has no web version, and the mobile app is not available in all countries.

Pocket is a website that allows its visitors to save news from the web for later reading. Despite the variety of sources, it is not possible to choose only trusted ones and the user has to decide for himself what deserves his attention and what does not. Pocket offers a summary of saved content, but only of individual articles.

2.2 Text Summarization

A number of text summarization methods have been developed in recent decades. They can be generally classified into extractive and abstractive [3].

The first type seeks to extract the key information from the text. It is the more traditional summarization method because of the simpler task it performs, namely returning a subset of the initial data. Abstractive methods, in turn, generate a completely new text, carrying the meaning of the original. Obviously, the task becomes more complex as there is much more freedom, but also the summaries generated can be in a much more natural format.

Each of these two types of methods has specific advantages and disadvantages. In extractive methods, each sentence is first evaluated for the amount of information it carries, and on that basis only the more meaningful sentences are left. This way, false facts cannot get into the summary, but often the structure of the text is not well logically connected and there may even be redundant sentences – if two sentences are very similar,

---

[1] https://about.smartnews.com/en/
[2] https://getpocket.com/home

the algorithm would choose them, even though they carry the same value. Abstractive methods, in turn, "understand" the content of the source data and generate new sentences summarizing the main points. They often use encoder-decoder architectures [4], such as transformative models. Their results look much more consistent, with preserved logical connections between sentences. The downsides are that this is a much more complex task and requires much larger databases to train the model. The text could contain grammatical errors or inaccuracies if the model is not well trained, there is also a danger of the biggest problem when we want to summarize material containing correct facts – the possibility of hallucinations, i.e. the model to generate content that does not exist in the source text.

Evaluating the performance of text summarization algorithms is not an easy task because there may be multiple valid abstracts for a single text. Obviously, no direct comparison can be made, and even a solution given by an experienced person will perform poorly on this test because everyone has their own writing style. One of the most commonly used metrics is ROUGE (Recall-Oriented Understudy for Gisting Evaluation) with several variations – ROUGE-N, ROUGE-L, ROUGE-S [5]. The main idea behind it is a comparison between a model-generated summary and reference summaries of the same text, usually created by humans. The advantage of the ROUGE metric is its similarity to human evaluation. It is easy to calculate and the logic behind it is understandable. Moreover, it is not dependent on the language of the summarized text. The disadvantages, on the other hand, are that the semantic meaning is not taken into account, the reference texts have a determining role, which may not be appropriate, and their length will determine the final evaluation. In all cases, a reliable evaluation of the quality of an abstract of multiple articles is possible only with the participation at certain stages of a human expert.

For the purposes of this study, experiments were done with several popular platforms providing means for automatic text summarization: Sassbook[3], QuillBot[4], Resoomer[5], Summarizer[6]

The Sassbook platform provides options for choosing the type of method (extractive or abstractive) to generate a summary of the given text and the degree of its compression. The assessment by human experts of the summaries created by the platform using an extractive method indicates that these summaries are logically consistent, contain a condensed version of the most important information in the text, but are often incomplete – they do not mention important points of the source text. Summaries created by Sassbook using an abstractive method have similar characteristics. It is noteworthy that some of them contain phrases that are taken directly from the original text without being reformulated – something that is not characteristic of abstractive methods in principle.

The QuillBot and Resoomer platforms are set to summarize text using an extractive method, and the Summarizer – using an abstractive method. Human expert evaluation shows that all three cope well with the task of summarizing a single text, with the generated summary being logically consistent, comprehensive and containing well-chosen information. None of the platforms studied provide capabilities for summarizing multiple

---

[3] https://sassbook.com/ai-summarizer
[4] https://quillbot.com/summarize
[5] https://resoomer.com/en/
[6] https://www.summarizer.org/

separately given texts, although both extractive and abstractive methods are applicable for the purpose.

2.3 Prompt Engineering

Prompt Engineering [6] is the engineering of specific input prompts to generative models that direct them to produce desired outputs. The prompt is a text in natural language, describing in detail the task that is being assigned. This technique can improve performance, especially on complex tasks.

BART does not require specific summarization prompts due to its architecture and the way it is trained. However, using them would be beneficial to improve the quality and relevance of the summaries generated. With GPT models, which are capable of performing a larger set of tasks, however, this is very important. The key techniques in constructing a good prompt fall into several categories. The first is instructional prompts, which can be both direct instructions, such as: "Summarize the following text: [input text]", and instructions on what role the model should step into to perform the task well; a specific profession or expertise is usually mentioned, for example: "You are a doctor. Analyze the symptoms described in the text: [input text]". Another technique is to add context. This can be any information that helps the model to understand the purpose, for example: "The following article contains information about climate change. Summarize the main points of it: [input text]". Asking questions also provides good guidance for eliciting relevant information: "What are the key points in the text: [text]?". For more complex tasks, they can be decomposed into smaller ones that are easier to work with, for example: "First extract the main points from the text, then summarize them in one paragraph.".

The most important thing is that the prompt be clear and unambiguous. As more specific the instructions are, there is a lower risk of a vague answer. Adding the necessary context directs the model. The best model results are usually achieved through a series of experiments with different prompts and improving in correspondence with the input data.

3 Data Collection and Preprocessing

3.1 Data Sources and Preprocessing Algorithms

Our approach is oriented towards working with publicly available articles from well-known news agencies such as CNN[7], BBC[8], Associated Press[9]. Data has been collected using the Playwright open source Python library[10], recording the title, content, author and date. The database is updated daily with the latest news. In order for the summary to present the information without changing the facts, certain words describing the time when the action takes place such as "today", "yesterday", "on Monday", etc. are replaced

---

[7] https://edition.cnn.com/
[8] https://www.bbc.com/
[9] https://apnews.com/
[10] https://scrapingrobot.com/blog/playwright-web-scraping/

with the corresponding dates. In the process of collecting the publications, keywords are extracted and recorded from them. Thus, they are ready to be indexed and categorized and the search of similar articles is expected to be easier later on.

Keyword extraction is a fundamental task in natural language processing that extracts the most relevant phrases from text. One popular method for solving this task is TF-IDF (Term Frequency-Inverse Document Frequency) [7]. The idea behind it is that if a word appears frequently in a document, but appears less frequently in several other documents in the corpus, then that word is likely to be important to that document. Conversely, if a word appears frequently in many documents, it is probably not very important for distinguishing them. Another popular method for the purpose is RAKE [8], it is particularly effective for extracting multi-word phrases that are important to the topic of a particular document. A combination of these methods usually gives the best result.

After extracting keywords by combining the described techniques, they should be normalized by lemmatization [9], a process of extracting the basic form of the word, then removing the repetitions. For a more in-depth analysis, word embeddings [10] are used, where words are represented as vectors of numbers and so it is easy to establish dependencies between them based on the context in which they are used, with similar meaning of nearby vectors. The standard cosine similarity can be used to compare vectors. It is necessary to define a threshold of similarity above which we consider two words to be synonyms. This way we can filter the keywords with the same meaning. After using these techniques, we will have the minimum number of keywords for each article and they will be stored for further use.

3.2 Technologies and Libraries Used

The article collection is done through the Playwright library. Designed as a library for automated testing, it is a good choice for the specific task because it can control the browser and handle content dynamically. Unlike traditional web scraping libraries that directly extract the HTML content, this one interacts with the web page as a real user would, thus solving problems with sites that actively use JavaScript.

Two separate databases have been selected for data storage. Pinecone DB[11] is chosen for word embeddings because it is designed specifically for storing vectors and, accordingly, for searching for vector similarity. This makes it very convenient for the purpose of the study, namely for semantic search for an answer to a question in news texts.

For the rest of the article information, MongoDB[12] is used, which is more traditional and optimized for searching unstructured data. It is needed so that we can finally display a list of links to the sources used for the summary.

The Python library SpaCy[13] has been used for keyword extraction. It has proven to be quite effective for natural language processing tasks. It is built on advanced machine learning algorithms and offers pre-trained models for many languages. It also excels at processing large volumes of text, making it a good choice for real-time applications like

---

[11] https://www.pinecone.io/
[12] https://www.mongodb.com/
[13] https://spacy.io/

the discussed one. In this case, we use it for tokenization, named entity recognition, and lemmatization.

Another key library suitable for use in implementing the proposed approach is Transformers[14]. It is a multifunctional natural language processing tool using the most modern models – transformers. It includes implementations of various transformer architectures such as BERT, GPT, T5, RoBERTa, XLNet, etc., and also provides access to a large collection of pre-trained models that can be fine-tuned for specific tasks. These models are trained on large volumes of data and achieve state-of-the-art performance on various benchmarks. A tokenizer and model are used in our case from Transformers, respectively BART in the case.

Another Python library in use is OpenAI[15], which provides the GPT-3.5 Turbo model[16]. When saving articles in Pinecone, it is necessary to transform them from text to vector. This is done using the SentenceTransformers library[17], and specifically the all-MiniLM-L6-v2 model[18], which transforms sentences into a 384-dimensional dense vector space and can be used for tasks such as clustering and semantic search.

4 Algorithms Used for Text Summarization

In the field of natural language processing, the task of summarizing multiple articles is a source of many challenges, such as preserving contextual relevance and logical structure. For the purpose of the discussed study, two of the most modern models have been selected – BART and GPT 3.5[19].

BART (Bidirectional and Auto-Regressive Transformer) [1] is a denoising autoencoder that uses standard sequence-to-sequence architecture with bi-directional encoder and left-to-right decoder. It is characterized by the fact that, thanks to bidirectionality, it can learn the relationships between words in both directions of a given sentence. It is trained on a large amount of data, allowing for deep language understanding. This makes it very flexible and usable for a wide range of tasks in the field of natural language processing. BART is particularly effective when fine-tuned for text generation, but it also works well for text understanding tasks. Thus, it achieves impressive results in answering questions, translating, abstract dialogue and summarizing. For the purpose of the presented study, the summarization pre-trained BartForConditionalGeneration was used, which was fine-tuned on a large collection of text – summary pairs of CNN and Daily Mail articles, facebook/bart-large-cnn[20]. This makes it very suitable for use in solving our task.

GPT-3.5 Turbo, on the other hand, is a purely autoregressive model, generating text token by token from left to right. All GPT models are designed to process and generate

---

[14] https://huggingface.co/facebook/bart-large-cnn, https://huggingface.co/docs/transformers/en/index
[15] https://github.com/openai/openai-python
[16] https://platform.openai.com/docs/models/gpt-3-5-turbo
[17] https://www.sbert.net/
[18] https://huggingface.co/sentence-transformers/all-MiniLM-L6-v2
[19] https://learn.microsoft.com/en-us/azure/ai-services/openai/concepts/models#gpt-35
[20] https://huggingface.co/facebook/bart-large-cnn

long sequences of text while preserving context and sequence between passages. They are trained on a large set of data and are very flexible, which allows them to perform a wide range of tasks beyond generalization, such as translation, question answering, leading a conversation.

In the course of the study, BERT (Bidirectional Encoder Representations from Transformers) [11], was also considered. It provides a bidirectional encoder, as well as BART, which helps to evaluate the context of a word against the words before and after it. However, due to its architecture, it cannot generate text, but can only recognize the sentences that carry the most meaning and value. This often does not result in well-ordered and logically connected text, and so BERT was not chosen as an option for text summarization, but it did an excellent job of filtering out synonyms when extracting the keywords.

Both models used to summarize long text have their strengths. In particular, BART is very robust to input deficiencies and very good at tasks requiring comprehension and reconstruction of text such as summarization. GPT-3.5 Turbo benefits from the large context it works with, making it more suitable for processing large volumes of text than many other models.

5 Software Implementation

The developed experimental software system consists of several main components that communicate with each other, as can be seen in Fig. 1. The user interface accepts the search query entered by the user, sends it to the programing interface that connects to the two databases, finds the requested information, processes it and returns the result to the frontend, which in turn visualizes it.

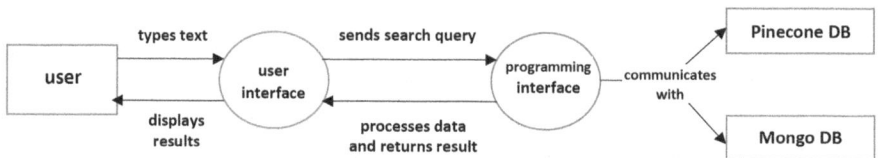

Fig. 1. System architecture.

5.1 Data Retrieval and Storage

The collection of data from the CNN, BBC and Associated Press sources is based on the HTML structure of the web pages, which is different for each of them. There are different approaches to choosing selectors against what is intended in the code. The most popular way is through CSS classes, but it is not always applicable because these days they are often dynamic and cannot be relied upon. Other options are through identifiers, data attributes, XPath (XML-based paths), etc. In the case of CNN and the Associated Press, the best option was to use CSS selectors, and the BBC's implementation suggests taking data attributes.

Thus, we provide the data in raw form, but for subsequent operations, it must undergo processing, described in more detail in Subsect. 3.1.

5.2 Programming Interface

The core of the system is implemented as a web page where the user can enter a keyword, sentence or question related to a particular topic. After pressing the search button, a POST request is sent to the Python API, in which the input of the user is directly submitted. Conversion of the input data to word embedding is performed using the all-MiniLM-L6-v2 model. Thus, it is in the same format as the articles stored in the Pinecone database.

The next step is a search on the corresponding vector, with the top_k parameter specifying the number of results we want to get. The list returned as a result of the operation also specifies a similarity score against the input data. In order to be able to consider only truly relevant information, it is filtered by a score greater than 0.5. The IDs of those records that match the given condition are then retrieved and searched in the MongoDB database where the rest of the article information is stored. From there, the texts, the keywords, on which similar articles can be suggested, as well as the links to the used articles, which would be useful in case the user wishes to make a reference, are obtained.

Having everything needed, it is possible to proceed to the main part – the summary. In addition to the basic functionality, the user has also supplementary optional settings available, such as filtering the included content by date/time period, etc. In addition, he/she can choose a specific summarization method in case of interest in it.

After retrieving and filtering the articles, it is proceeded to the selection of a model for summarization according to the set parameters and its recall. In the current implementation, a BART model pre-trained on news data for summarization is used by the Transformers library. To get better results, various parameters are set in the functions provided by the library. The first step is encoding by the tokenizer, which is converting the input text into tokens that the model can understand. The maximum length of the tokenized input data, in this case 1024, is also passed there to ensure that the model constraints are met.

This would be enough to summarize a single article, but here we have a set of articles and the so-called hierarchical summarization presented in Sect. 6 is required. The last step is decoding the returned tokens from the previous stage into text.

The other model provided to the user as a choice is the GPT-3.5 Turbo, which currently emerges as the optimal option as a combination of speed, price and performance. It is worked with through the so-called Chat Completions API[21] – the model takes a list of messages as input and returns the generated response as output. The list of incoming messages is actually an array of objects, each of which has a content and a role, which can be as follows: system, user, or assistant. By default, a conversation is formed with a single system message initially, followed by alternating user and assistant messages. This first message sets the assistant's behavior. This is when the model prompt discussed in Subsect. 2.3 takes place. For example, the personality of the assistant can be determined – whether to be an expert in a certain field or to give specific instructions on how to behave during the conversation. This system message is optional and if not specified will default to "You are a helpful assistant".

---

[21] https://platform.openai.com/docs/api-reference/chat

In the discussed implementation, the role of the assistant is chosen as "You are an experienced news analyst who is tasked with summarizing news articles based strictly on the given text, without adding extraneous information". This way we can ensure that the model will not provide data that is absent in the original text, i.e. it will not invent information.

The user message, in turn, reflects the request that the assistant must satisfy. In the current situation, this is a request to summarize the corresponding text which represents the content of several articles. This can happen in a number of ways, as discussed in Sect. 6.

The third type of messages that the OpenAI API accepts are those to the assistant. They usually store its previous responses, but can be submitted, their purpose being to serve as an example of the desired behavior. In the present case, since summarization is not an unambiguous task and an example could even confuse the model, this step is omitted.

All already described functionalities of the system are available both through the user and through the programming interface, which makes the implementation flexible and allows integration with other systems.

5.3 User Interface

The user interface offers an input field and a submit query button that becomes active and changes color when a search query is entered. When indicating the inscription "Advanced options" or additional settings, the cursor changes, which signals that this is also a button, and when pressed, we will see more options. The web application contains a single page that doesn't reload, and the content changes via JavaScript based on events like button presses, etc.

Figure 2 shows what the user interface looks like with the additional settings menu open and the search button active.

The combination of searching for the information in the two databases and the summarization of several articles takes time to process, which could confuse the user and make him/her enter the same query multiple times. To avoid this, a visual indication of the process has been added in the form of an hourglass, which rotates while the query is executed and disappears when the result is displayed.

After the server returns the summary and links to the articles used, they are visualized as shown in Fig. 3. The selected model in this case is GPT-3.5 Turbo.

What a summary of the same articles generated with the BART model looks like can be seen in Fig. 4.

Thanks to CSS media queries[22], the GUI looks good on different devices and resolutions, especially on mobile resolution.

---

[22] https://www.w3schools.com/css/css_rwd_mediaqueries.asp

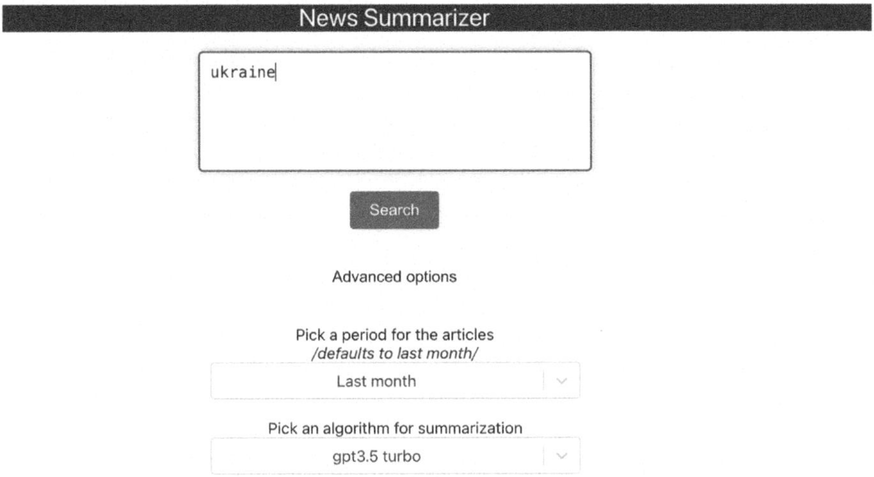

Fig. 2. Additional settings in the user interface.

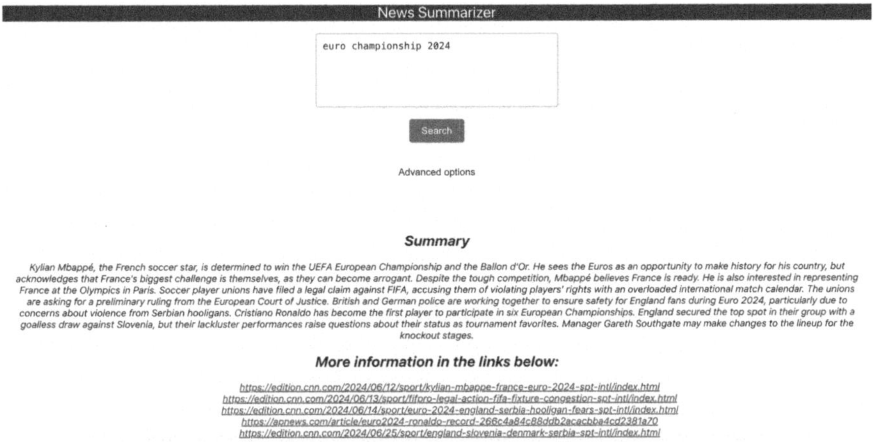

Fig. 3. Result when searching for information through the user interface with the GPT-3.5 Turbo model selected.

6 Analysis of Conducted Experiments

Summarizing multiple articles into one complete and comprehensive abstract is a challenging task that requires both accuracy and coherence. Recent improvements in transformer-based models are quite promising for the solution of this task. In the development process, a large number of experiments were done to evaluate and compare the performance of different models.

The approach of summarizing several articles is key to the final result. All models were experimented with three schemes: sequential summarization, hierarchical summarization, direct summarization.

Fig. 4. Same search result as in Fig. 3 but BART model selected.

Sequential summarization. Each article is summarized individually, and the individual abstracts are then combined and further summarized to produce a single overall summary.

Hierarchical summarization. Articles are divided into clusters based on their content. Each cluster is summarized and the summarized clusters are combined into a final summary.

Direct summarization. All articles are merged from the beginning, and the model generates a summary of the entire content without further processing.

In sequential summation, both BART and GPT-3.5 Turbo perform well on individual summaries. BART's results, however, sometimes seem to lack a logical connection between individual sentences, as they contain only the most important information from each article, unlike those generated by GPT, which also contain some of the important details. In hierarchical summarization, the BART output is much more structured, but again GPT does a better job of balancing detail and retelling consistency.

In the direct approach, BART runs into input size limits, so the text of the articles is truncated to 1024 tokens, and so the key points of all the news are not present in the summary. With GPT, this problem also occurs, but only with a combination of quite long articles, due to the significantly more acceptable model constraints. 1024 tokens, which is the limit on BART, equals about 768 words on average, while 4096 tokens on GPT-3.5 Turbo equals about 3072 words.

To deal with the model limits mentioned, several options were considered for dividing the text into segments to be processed separately [12]:

- *Fixed-length chunks:* dividing the text into equal chunks (for example of 1024 tokens). This approach is easy to understand and implement. All parts are the same size, which also simplifies processing. The negatives are that a fixed length may require cutting off sentences or paragraphs, thus losing context and consistency.
- *Sentence-based segmentation:* the text is split into parts containing whole sentences, ensuring that each part does not exceed the limit. With this approach, sentence boundaries are preserved and thus context and readability are not lost. However, the sizes of the pieces can vary, leading to inefficiencies if some pieces are much shorter than

others. Additionally, sentence boundary detection gives additional complexity to the task.
- *Paragraph-based segmentation:* split into parts based on the paragraphs in the text, with each part not exceeding the model limit. This way preserves logical connection and context, because paragraphs themselves are often complete logical units, and it makes summaries more coherent. The negatives are similar to those of sentence segmentation, but in addition, the method may be inapplicable if paragraph boundaries were removed during initial data collection.
- *Sliding window method:* split the text into overlapping parts. This method ensures that any important information will be preserved in the individual pieces, and greatly reduces the risk of losing valuable context. However, this leads to redundant processing of the same text, which increases complexity.
- *Thematic partitioning (clustering):* using additional algorithms, to divide the text into parts based on the topics covered in them. This approach maintains thematic coherence, making abstracts more meaningful. However, it requires additional processing with topic detection and segmentation algorithms. As with other methods, here individual parts can also vary significantly in volume.

Each of these methods has its advantages and disadvantages. Choosing the right approach depends on the specific requirements of the task and the nature of the text. Splitting into fixed-length chunks is easy, but can lead to context loss. Paragraph-based segmentation preserves logical relationships, but was not applicable because of the way articles are saved in the database. The sliding window method also does not allow loss of information, but it leads to redundant operations, which is not desirable for real-time software, such as our case. Thematic clustering also complicates the process a lot and increases the user's waiting time. Therefore, the approach chosen was sentence-based segmentation.

Fine-tuning of GPT-3.5 Turbo includes experimenting with p words for different prompts. Different expert roles were set for the model, as well as different guidelines and formats for the generated text – how detailed it should be, what parts it should contain (for example, introduction, body, conclusion or key points), etc.

When generating a summary with BART, various parameters controlling the performance of the model can be set, such as: the number of token sequences in a beam search during text generation, the minimum/maximum length of the generated sequence, etc. To determine the size of the summary generated by BART, the current implementation uses a function that calculates the minimum and maximum length based on the length of the input text and the specified compression ratio.

The experimental results highlight the strengths and weaknesses of BART and GPT-3.5 Turbo in the multi-article summarization task. While BART is notable for robust performance and structured summaries, GPT features text coherence, better readability, and more complete coverage of key content points. The choice of summarization strategy significantly affects the performance of both models, with hierarchical summarization proving to be the most effective approach.

These findings highlight the potential of models based on transformer architectures in summarization tasks and provide valuable directions for future research and practical implementations in the field of natural language processing.

The discussed experiments were mainly aimed at investigating the effectiveness of the proposed approach and the possible alternatives it supports. Experiments are currently being planned to evaluate its performance in real-time operation.

7 Conclusion

The goal of the presented research – to create software gathering news from trusted media, which will satisfy a given user enquiry with concise, reliable information concerning a certain period of time, has been successfully fulfilled. Functionality is offered that is accessible through both a user interface and a programing interface. The interfaces also provide additional options such as filtering articles by period and selecting a summary model. Access to the original articles is provided for reference as well as a set of extracted keywords that can be used for subsequent searches. The algorithms and technologies used provide an opportunity to scale the project. The successful application of two models – BART and GPT-3.5 Turbo – in summarizing multiple news articles confirms the capabilities of these language models as aids in processing and understanding large volumes of text.

The results discussed in the paper make a practical contribution to the field of natural language processing by developing and experimenting a reliable approach to using a large language model such as GPT to generate a content-correct summary of multiple texts, avoiding the inherent danger of model hallucinations.

Our future plans include adding more models, which will no doubt be coming, and also expanding the capabilities of the current ones by being trained in multiple languages, which will make the tool more useful on a global scale. The user interface could be improved to allow the users to adjust the model prompt by giving them the freedom to choose the format of the summary, whether it should be presented as a bulleted list or in logically linked text, which be the focus and how detailed and long it should be. Another functionality that could be added is text-to-speech conversion, which would allow users to listen to summarized articles while doing other activities.

Acknowledgments. This study is financed by the European Union-NextGenerationEU, through the National Recovery and Resilience Plan of the Republic of Bulgaria, project No BG-RRP-2.004-0008.

Disclosure of Interests. The authors have no competing interests to declare that are relevant to the content of this article.

References

1. Lewis, M., Liu, Y., Goyal, N., et al.: BART: denoising sequence-to-sequence pre-training for natural language generation, translation, and comprehension. In: Proceedings of the 58th Annual Meeting of the Association for Computational Linguistics (online, 5–10 July 2020), pp. 7871–7880. Association for Computational Linguistics, Stroudsburg, PA, USA (2020)
2. Kalyan, K.: A survey of GPT-3 family large language models including ChatGPT and GPT-4. Nat. Lang. Process. J. **6**, 100048 (2024)

3. Yadav, D., Desai, J., Yadav, A.: Automatic text summarization methods: a comprehensive review. arXiv:2204.01849 [cs.CL], 3 March 2022. https://arxiv.org/pdf/2204.01849. Accessed 20 Aug 2024
4. Rahul, S.: Encoder Decoder Architecture. Medium, 12 March 2024. https://ogre51.medium.com/encoder-decoder-architecture-ca673109243f. Accessed 20 Aug 2024
5. Steinberger, J., Ježek, K.: Evaluation measures for text summarization. Comput. Inform. **28**, 1001–1026 (2009)
6. Ziegler, A., Berryman, J.: A developer's guide to prompt engineering and LLMs. The GitHub Blog, 17 July 2023. https://github.blog/2023-07-17-prompt-engineering-guide-generative-ai-llms/. Accessed 20 Aug 2024
7. Simha, A.: Understanding TF-IDF for Machine Learning: A gentle introduction to term frequency-inverse document frequency. CapitalOne, 6 October 2021. https://www.capitalone.com/tech/machine-learning/understanding-tf-idf/. Accessed 20 Aug 2024
8. Naskar, A.: Keyword Extraction using Rake algorithm in Python. Medium, 23 August 2023. https://thinkinfi.medium.com/keyword-extraction-using-rake-algorithm-in-python-be29595af30. Accessed 20 Aug 2024
9. Murel, J., Kavlakoglu, E.: What are stemming and lemmatization? IBM, 10 December 2023. https://www.ibm.com/topics/stemming-lemmatization. Accessed 20 Aug 2024
10. Brownlee, J.: What are Word Embeddings for Text? Machine Learning Mastery, 7 August 2019. https://machinelearningmastery.com/what-are-word-embeddings/. Accessed 20 Aug 2024
11. Horev, R.: BERT Explained: State of the art language model for NLP. Medium, 10 November 2018. https://towardsdatascience.com/bert-explained-state-of-the-art-language-model-for-nlp-f8b21a9b6270. Accessed 20 Aug 2024
12. Jain, M.: RAG: Part 2: Chunking. Medium, 5 April 2024. https://medium.com/@j13mehul/rag-part-2-chunking-8b68006eefc1. Accessed 20 Aug 2024

Reinforcement Learning Control of Cart Pole System with Spike Timing Neural Network Actor-Critic Architecture

Borislav Markov(✉) and Petia Koprinkova-Hristova

Institute of Information and Communication Technologies, Bulgarian Academy of Sciences, Sofia, Bulgaria
{borisslav.markov,petia.koprinkova}@iict.bas.bg

Abstract. This work presents a biologically plausible approach to reinforcement learning control of benchmark cart-pole system with continuous state and discrete actions. The proposed solution utilizes brain-inspired spiking neural networks (SNN), constructed with the aid of the NEST simulator in Python. To update the SNN connection weights, the Temporal Difference TD(0) algorithm is combined with Spike Timing Dependent Plasticity (STDP), reflecting principles of reinforcement learning. It is shown that SNN actor-critic architecture solves the task several times faster than classical neural networks.

Keywords: Spike timing neural network · Spike timing dependent plasticity · Reinforcement learning · Dopamine · Cart Pole system

1 Introduction

Within RL theory, the aim is to discover a policy (a mapping from sensory input to motor actions) that maximizes the anticipated cumulative future reward. The book from Sutton and Barto [18] has become a foundational resource in the field of reinforcement learning. Temporal difference (TD) learning algorithms are widely used in various reinforcement learning (RL) applications, including game playing [1], robotics, and finance, due to their efficiency and ability to learn directly from interactions with the environment.

In the realm of neuroscience, two phenomena, Spike Timing-Dependent Plasticity (STDP) and the neurotransmitter dopamine, have long captured the attention of researchers [3,20,21].

STDP stands as a cornerstone of synaptic plasticity, reflecting the brain's remarkable feature to modify the strength of connections between neurons based on the precise timing of their activity. This form of Hebbian plasticity, initially proposed by Donald Hebb [8] in 1949, posits that synapses are strengthened when the presynaptic neuron consistently fires just before the postsynaptic neuron, and weakened when the firing order is reverse. STDP has emerged as a fundamental mechanism underlying learning and memory formation in various brain regions.

The understanding of the basal ganglia's function was primarily shaped by a sequence of experiments conducted by Schultz and his colleagues [15], focusing on the activity of midbrain dopamine neurons in primates. During a conditioned reaching task, dopamine neurons initially react to the reward following successful trials. However, as the animal becomes proficient in the task, dopamine neurons begin to react to the conditioned stimulus and no longer respond to the actual delivery of the reward. This suggests that dopamine neuron activity not only reflects immediate rewards but also anticipates future rewards.

A schematic diagram was proposed by Kenji Doya in his work [5], outlining reward-based learning within the basal ganglia. In this model, he explores possible connections within the cortico-basal ganglia loop and the potential functions of its components in a reinforcement learning framework. Specifically, Doya suggests that neurons in the striatum forecast the future reward associated with the current state and potential actions. Sutton and Barto [18] showed hypothetical neural implementation of an actor-critic structure where the actor and the value-learning part of the critic are respectively placed in the ventral and dorsal subdivisions of the striatum.

There are few works exploiting biologically plausible spike timing neural networks with RL framework. Tieck et al. [19] demonstrate how to control a robotic arm using reinforcement learning with dopamine modulated STDP. They used dopamine dependent STDP synapses for connections between sensory and action populations.

Shim and Li [16] introduce a technique for reinforcement learning in mobile robot collision avoidance scenarios. Their approach involves the design of novel reward functions considering factors such as the distance to obstacles, distance to the target, and other relevant parameters. They investigate the concept of Reward-Modulated Spike Timing Dependent Plasticity (RM-STDP), suggesting that SNNs could learn an optimal decision-making policy. This learning occurs under a carefully crafted global reward signal with an eligibility trace mechanism, facilitating the handling of delayed rewards.

Jitsev et al. [9–11] extend a recent spiking actor-critic network model of the basal ganglia, aiming to create a minimal realistic model of learning from both positive and negative rewards. Spiking network is implemented with NEST Simulator and experiments involve applications of delayed reward on a task of moving an agent in a grid world trying to maximize the reward. They demonstrate that the network can learn not only to approach the positive rewards, but also to avoid punishments.

Inspired by the theories of brain structures involved in RL in this work we proposed an SNN actor-critic architecture whose STDP synapses were trained by dopamine signal proportional to the TD error. Its work was tested on the benchmark task for cart-pole [2] control. The main contributions are as follows:

- An approach for converting of the continuous state vector of the object (cart-pole) to a strictly positive input to the actor-critic SNN was proposed.
- Actor was designed as a winner-take-all (WTA) SNN structure.

NEST Simulator library [7] was used throughout the experiments.

2 Methods

2.1 TD Learning and Actor-Critic Design

The Actor-Critic is a reinforcement learning architecture that combines elements of both value-based and policy-based RL methods. The critic aim is to predict the discounted sum of future rewards:

$$G(t) = \sum_{k=0}^{\infty} \gamma^k r(t+k+1) \tag{1}$$

where $r(t+k+1)$ is the reward at step $t+k+1$.

The state value function of a given policy π that should be predicted by the critic is then defined as:

$$v_\pi(s) = \mathbb{E}_\pi[\sum_{k=0}^{\infty} \gamma^t r(t+k+1) | s(t) = s], \forall s \in \mathcal{S} \tag{2}$$

where $\mathbb{E}$ denotes the expectation operator, $v_\pi(s)$ is the value function for state s and policy π, $\mathcal{S}$ is the set of all possible states. For a given state $s(t)$ the above equation can be rewritten in the recursive form:

$$v_\pi(s(t)) = \mathbb{E}_\pi[r(t+1) + \gamma v_\pi(s(t+1))] \tag{3}$$

The policy is improved using critic value predictions as follows:

$$\pi'(s) = \mathrm{argmax}_{a \in \mathcal{A}} \mathbb{E}[r(t+1) + \gamma v_\pi(s(t+1)) | s(t) = s, a(t) = a] \tag{4}$$

TD learning, as proposed by [17], is an incremental learning approach that uses past experience to predict future behaviour of a completely unknown system via credit assignment by means of difference between two successive predictions called TD error. Particularly, in RL framework TD error represents the difference between the actual estimate of the value function $\hat{v}_\pi(s(t))$ by the critic and its target value $(r(t+1) + \gamma \hat{v}_\pi(s(t+1)))$ from Eq. (3):

$$\delta(t) = r(t+1) + \gamma \hat{v}(t+1) - \hat{v}(t) \tag{5}$$

The TD error (5) plays a vital role in both value function learning and policy formation within an actor-critic framework. From the view point of neurobiology its is associated with dopamine neurons activity.

2.2 The Cart Pole Object

The cart-pole system is a classic control problem for balancing an inverted pendulum (the pole) on top of a moving cart (Fig. 1). It is often used as a benchmark to test and develop algorithms [13] in control theory and reinforcement learning. The system consists of a cart that can move horizontally along a track and a pole that is attached to the cart by a hinge. The goal is to balance the pole upright by moving the cart left or right. A reward of +1 is given for every step including the termination step. Episode finishes when the pole falls down. The system state and reward signals originating from the cart-pole environment are modified to directly stimulate the SNN actor-critic neural network, facilitating its spiking activity.

Observation Space is a 4-dimensional vector

$$S_o = [d, v, \theta, \dot{\theta}]^T, \tag{6}$$

representing the state of the system, including the cart position($d \in [-4.8, 4.8]$), cart velocity($v \in [-\infty, \infty]$), pole angle ($\theta, rad \in [-0.418, +0.418]$), and pole angular velocity ($\dot{\theta} \in [-\infty, \infty]$). Action Space is a discrete set of 2 actions representing pushing the cart left($a = 0$) or right($a = 1$).

When the pole angle goes beyond ± 0.13 radians, the pole is already falling. Since intuitively any control algorithm should try to keep the cart around the center, values of d beyond ± 1.5 should not be considered. Hence the cart velocity also should not be too high. The same is valid for pole angle and its angular velocity. Hence here we propose to clamp the values of state variables within following intervals: $S_{c,min} = \{-1.5, -1.5, -0.13, -2.1\}$ and $S_{c,max} = \{1.5, 1.5, 0.13, 2.1\}$.

Since our model accepts only positive input signals, in order to account for both sign and magnitude of the state vector, next step is to transform the clamped state vector into 8-dimensional one as follows:

$$\begin{array}{l} \text{IF } S_{c_i} > 0 \text{ THEN } S_{p_{i1}} = |S_{c_i}|, \quad S_{p_{i2}} = 0 \\ \quad\quad\quad \text{ELSE } \quad S_{p_{i1}} = 0, \quad S_{p_{i2}} = |S_{c_i}| \end{array} \tag{7}$$

In this way the information about the sign and magnitude of each element of the state vector S_{c_i} is coded in a two-dimensional vector $S_{p_i} = [S_{p_{i1}} \; S_{p_{i2}}]$ first element of which corresponds to the positive while the second one - to the negative value of the object state S_{o_i}. Finally the transformed state vector $S_p = [S_{p_1} \; S_{p_2} \; S_{p_3} \; S_{p_4}]$ is scaled within interval $[0.2, 1.5]$.

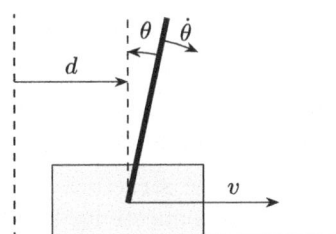

Fig. 1. The Cart Pole object. Observation vector is cart position, cart velocity, pole angle, and pole angular velocity denoted as $\{d, v, \theta, \dot{\theta}\}$. Action takes discrete values ($a = 0$ for moving left and $a = 1$ for moving right).

2.3 Structure of the Model

The structure of the actor-critic SNN is given on Fig. 2. Environment's observation vector S_o is first clamped and then transformed, according to the rule (7) to S_p, and then scaled. Then it is fed into the module 'Internal State' as a Poisson spike train ('PG' on figure) scaled by S_{p_i}, $i = 1 \div 8$ to the eight subgroups of spiking neurons (each one of eight neurons) shown as violet and pink squares on Fig. 2. In this way the group of 8×8 neurons denoted as 'Internal state' generates spikes with frequency proportional to the positive (pink) and negative (violet) values of the original object state. The elements 'Critic' and 'Actor' are two groups of 50 and 60 neurons respectively that receive the 'Internal state' spikes via synapses with spike timing dependent plasticity (STDP) as in structure proposed in [18]. The 'Critic' should predict the value (3) while the 'Action' should generate the action (4). The dopamine signal is generated by a group of 50 neurons 'DA' whose spiking frequency is proportional to the TD error (5). The reward signal was fed as another Poisson spike train ('Reward' on figure) proportional to the reward from the environment.

The detailed structure of 'Actor' is shown on Fig. 3. It was designed as a Winner Takes All (WTA) structure. The mechanism involves two competing groups of neurons A_{left} and A_{right}, corresponding to the two possible actions, that inhibit each other via an inhibitory group of neurons INH. Each group in this structure consists of 20 neurons. When one of the action groups becomes slightly more active than the other it suppresses the activity of its competitor. In this way only one action group spikes thus effectively "winning" the competition.

This approach prevents the network from reaching an equilibrium, particularly when one neuronal group exhibits slightly higher spiking activity than others. Hereby used WTA is an interpretation of the perspective presented by Gerstner and Kistler in their book [6] and also Morita, Jitsev & Abigail [12], wherein the direct connections between action groups are voided and a Poisson spiking generator ($Noise$) is incorporated. WTA is a soft-max action selection with option to control exploration-exploitation ratio.

Choosing of action at the end of each step in an episode is based on number of spikes n_{spikes} in action groups as follows:

$$a_{t_n} = \begin{cases} left, n_{spikes}(A_{left}, t_{n-1}, t_n) > n_{spikes}(A_{right}, t_{n-1}, t_n) \\ right, otherwise \end{cases}$$

For all neurons in the structure the leaky integrate-and-fire (LIF) spiking neuron model is used. Connectivity between all neuron groups from Fig. 2 and Fig. 3 are designed via 'all_to_all' rule except those of STDP synapses that are using 'pairwise_bernoulli' rule.

The programming realization of the Actor-Critic SNN is in Python using NEST Simulator library [7]. The Cart Pole is simulated using Gym environment [13]. Both modules interact on a request-response basis.

Figure 4 shows distribution of a single time step from t_n to t_{n+1}. It starts with a period of competition in the WTA group during which the winning action

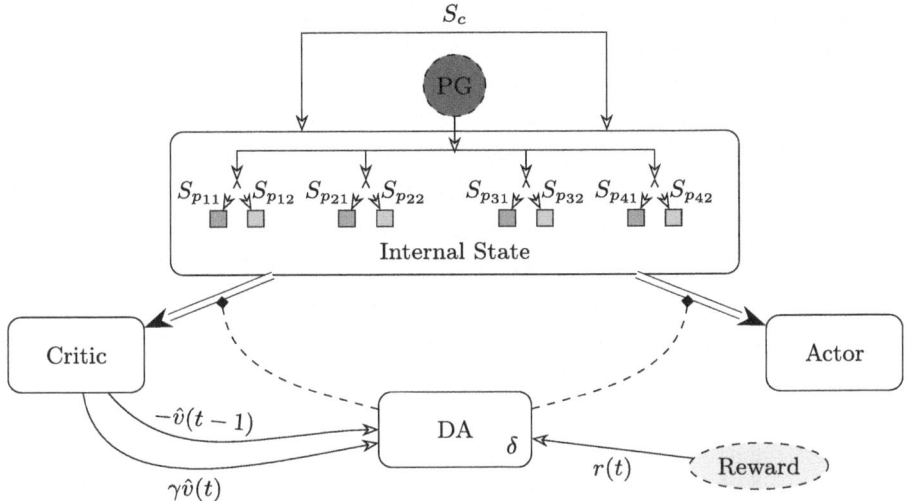

Fig. 2. Actor-Critic SNN network. Single-line arrows represent fixed weight connections, while double-line arrows denote dopamine-dependent synapses. Dashed lines denote the dopamine fed to the STDP synapses.

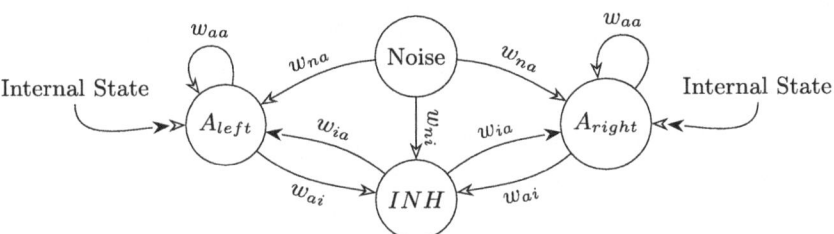

Fig. 3. The Actor. Detailed view of WTA connections. Filled arrows are inhibitory connections, opened arrows are excitatory connections. Parameters: $w_{aa} = 10.5$; $w_{na} = 2.1$; $w_{ni} = 0.9 w_{na}$; $w_{ia} = -2.6$; $w_{ai} = 2.8$.

has to be selected. Thereafter STDP learning is initiated followed by a period of relaxation. While WTA neurons are active as well as during the relaxation period STDP is blocked by feeding negative current of -100pA into DA group of neurons. During the last episode the relaxation time is prolonged ten times in order to settle down the SNN state.

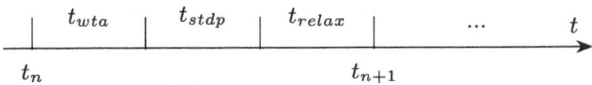

Fig. 4. Time Scale and break down of discrete steps t_n. Parameters: $t_{wta} = 50$ ms, $t_{stdp} = 20$ ms, $t_{relax} = 40$ ms. For the last step of an episode $t_{relax_{last}} = 10 * t_{relax}$ ms

3 Simulation Results

Due to randomness in the actor-critic SNN that is coming from several noise generators and random connectivity in the STDP synapses, not every learning trial finishes successfully. Nevertheless, successive trials achieved learning scores above 195 in the first 10 episodes which is much faster than any other solution known from literature [4,13].

The accumulated reward during a successful trial of 200 episodes is shown on Fig. 5.

Simulation is compared to the random-policy agent and outbeats it definitely.

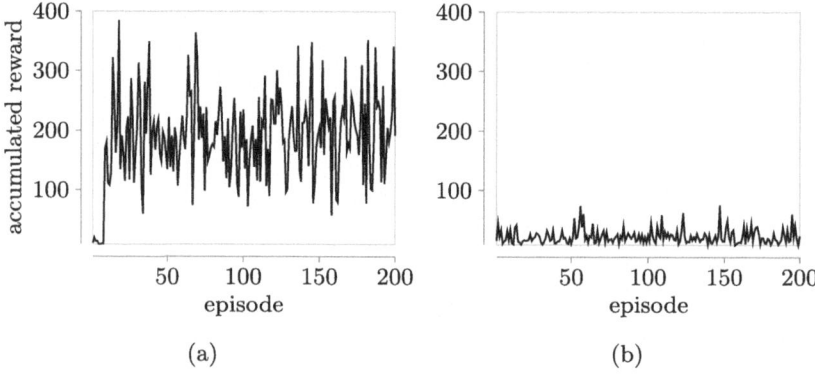

Fig. 5. Cumulative reward per episodes for: (a) SNN agent; (b) Random policy agent

The speed of the learning process can partially be influenced by changing the scaling of the reward, which at the moment is $r_{scale} = 6.25$. It controls the 'Reward' Poisson spike generator that feeds the group 'DA'. The value is empirically derived by experiments. Other parameters that can influence the learning speed are amplitudes A_+ and A_- of the weight changes in STDP learning rule [14]. In current work they were experimentally set to $A_- = 0$ and $A_+ = 0.004$. Since the environment continues giving a reward of +1 even though the pole is falling and episode will terminate soon, there is also a need to modify the time constants τ_n and τ_c, which influence both the dopamine and the eligibility trace respectively They were set experimentally to $\tau_n = 10$ and $\tau_c = 50$.

Figure 6 depicts a fragment of spikes for a period of 2 s for all neuronal groups during a chosen episode. Whole simulation lasts for $t_{Simulation} = n_{Episodes} * (n_{Steps} * (t_{wta} + t_{stdp} + t_{relax}) + 9 * t_{relax})$ milliseconds. For example in case of $n_{Episodes} = 200$ episodes with an average number of steps per episode $n_{Steps} = 200$ it would take approximately $200 * (200 * (50 + 20 + 40) + 360) = 4472000$ ms.

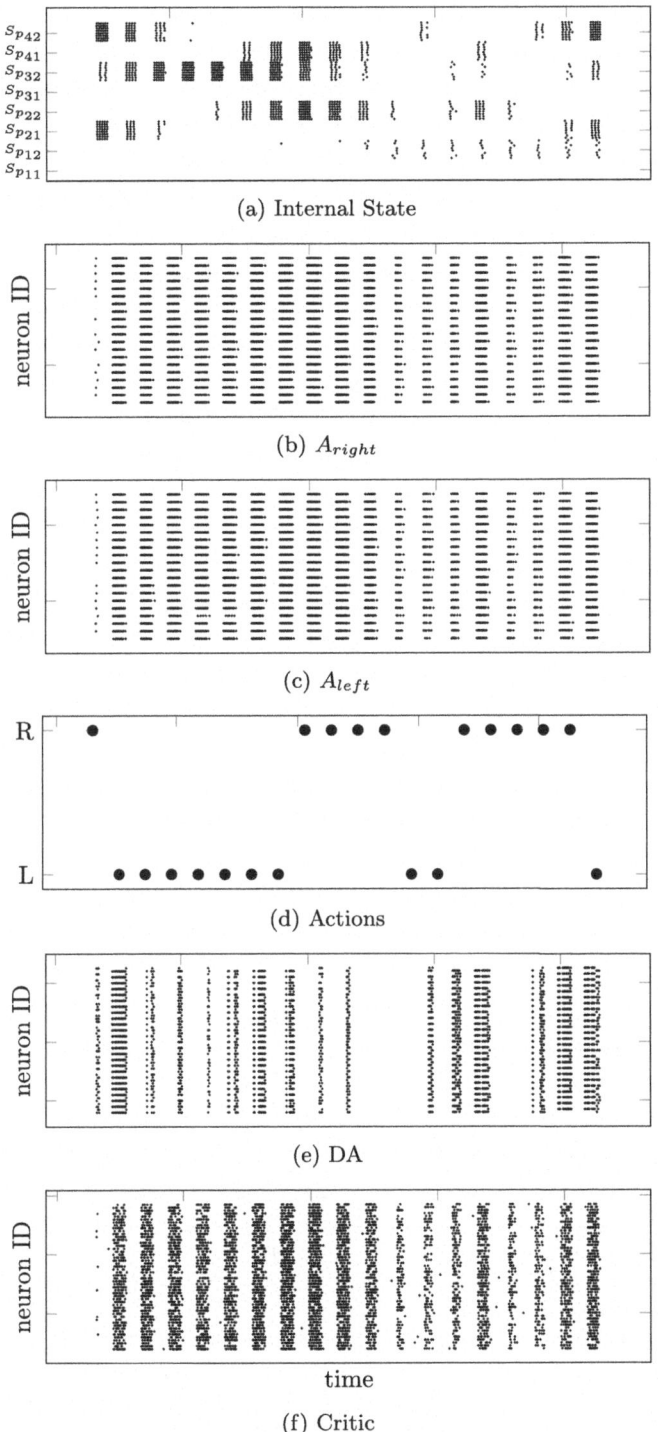

Fig. 6. Spikes of neuronal groups for two seconds during a chosen episode.

4 Conclusions

This work contributes to the few practical experiments already available in the field of reinforcement learning for motor control using SNNs. Primary goal is to demonstrate a minimal connectivity that is capable of learning and driving a mechanical part. Further work aims application of this approach to a practical task in robotics.

The overall SNN architecture could be further refined by incorporation of knowledge about basal ganglia and its role in reinforcement learning in the brain.

References

1. Barto, A.G.: Reinforcement learning: connections, surprises, and challenge. AI Mag. **40**(1), 3–15 (2019)
2. Brockman, G., et al.: Openai gym. arXiv preprint arXiv:1606.01540 (2016)
3. Chen, W.: Neural circuits provide insights into reward and aversion. Front. Neural Circ. **16**, 1002485 (2022)
4. Chevtchenko, S.F., Bethi, Y., Ludermir, T.B., Afshar, S.: A neuromorphic architecture for reinforcement learning from real-valued observations. arXiv preprint arXiv:2307.02947 (2023)
5. Doya, K.: Complementary roles of basal ganglia and cerebellum in learning and motor control. Curr. Opin. Neurobiol. **10**(6), 732–739 (2000)
6. Gerstner, W., Kistler, W.M., Naud, R., Paninski, L.: Neuronal Dynamics: From Single Neurons to Networks and Models of Cognition. Cambridge University Press (2014)
7. Gewaltig, M.O., Diesmann, M.: Nest (neural simulation tool). Scholarpedia **2**(4), 1430 (2007). https://doi.org/10.4249/scholarpedia.1430
8. Hebb, D.: The organization of behavior. Emphnew York (1949)
9. Jitsev, J., Abraham, N., Morrison, A., Tittgemeyer, M.: Learning from delayed reward und punishment in a spiking neural network model of basal ganglia with opposing D1/D2 plasticity. In: Villa, A.E.P., Duch, W., Érdi, P., Masulli, F., Palm, G. (eds.) ICANN 2012. LNCS, vol. 7552, pp. 459–466. Springer, Heidelberg (2012). https://doi.org/10.1007/978-3-642-33269-2_58
10. Jitsev, J., Abraham, N., Morrison, A., Tittgemeyer, M.: Functional role of opponent, dopamine modulated D1/D2 plasticity in reinforcement learning. BMC Neurosci. **14**, 1–2 (2013)
11. Jitsev, J., Morrison, A., Tittgemeyer, M.: Learning from positive and negative rewards in a spiking neural network model of basal ganglia. In: The 2012 International Joint Conference on Neural Networks (IJCNN), pp. 1–8. IEEE (2012)
12. Morita, K., Jitsev, J., Morrison, A.: Corticostriatal circuit mechanisms of value-based action selection: implementation of reinforcement learning algorithms and beyond. Behav. Brain Res. **311**, 110–121 (2016)
13. Nagendra, S., Podila, N., Ugarakhod, R., George, K.: Comparison of reinforcement learning algorithms applied to the cart-pole problem. In: 2017 International Conference on Advances in Computing, Communications and Informatics (ICACCI), pp. 26–32. IEEE (2017)
14. Potjans, W., Morrison, A., Diesmann, M.: Enabling functional neural circuit simulations with distributed computing of neuromodulated plasticity. Front. Comput. Neurosci. **4**, 141 (2010)

15. Schultz, W., Apicella, P., Ljungberg, T.: Responses of monkey dopamine neurons to reward and conditioned stimuli during successive steps of learning a delayed response task. J. Neurosci. **13**(3), 900–913 (1993)
16. Shim, M.S., Li, P.: Biologically inspired reinforcement learning for mobile robot collision avoidance. In: 2017 International Joint Conference on Neural Networks (IJCNN), pp. 3098–3105. IEEE (2017)
17. Sutton, R.S.: Learning to predict by the methods of temporal differences. Mach. Learn. **3**, 9–44 (1988)
18. Sutton, R.S., Barto, A.G.: Reinforcement Learning: An Introduction. MIT Press (2018)
19. Tieck, J.C.V., et al.: Learning target reaching motions with a robotic arm using dopamine modulated STDP. In: 18th IEEE International Conference on Cognitive Informatics and Cognitive Computing (2019)
20. Wang, J., Belatreche, A., Maguire, L., McGinnity, T.M.: An online supervised learning method for spiking neural networks with adaptive structure. Neurocomputing **144**, 526–536 (2014)
21. Yamazaki, K., Vo-Ho, V.K., Bulsara, D., Le, N.: Spiking neural networks and their applications: a review. Brain Sci. **12**(7), 863 (2022)

Predictive and Explainable Modelling in Economics on the Case Study of Remittance Prediction Using the NeuDen AI Computational Architecture

Iman AbouHassan[1,2](✉) [iD], Nikola Kasabov[1,3,4,5](✉) [iD], Roumen Trifonov[1] [iD], and George Popov[1] [iD]

[1] Technical University of Sofia, Sofia, Bulgaria
`iabouhassan@tu-sofia.bg, nkasabov@aut.ac.nz`
[2] Central Bank of Lebanon, Beirut, Lebanon
[3] KEDRI, SECMS, Auckland University of Technology, Auckland, New Zealand
[4] IICT, Bulgarian Academy of Sciences, Sofia, Bulgaria
[5] Dalian University, Dalian, China

Abstract. The presence of complex time series and multimodal data in economics and finance necessitates the development of advanced analytical models capable of interpreting complex patterns and dynamics. Despite their fast development as LLM, the current AI systems cannot do that. This paper introduces how a novel computational architecture, NeuDen, which combines the strengths of evolving spiking neural networks (eSNNs) with the interpretability of evolving dynamic neuro-fuzzy systems (eDNFS), can be used to model high-dimensional temporal data in economics. The NeuDen model is applied to the analysis of remittance inflows, demonstrating its ability to not only model data but also uncover the underlying trends in intricate remittance data. This application highlights the model's potential as a tool for efficient modelling of economic data that also helps understand the dynamics of economic processes.

Keywords: NeuDen · Evolving Spiking Neural Networks · Evolving Connectionist Systems · Evolving Hybrid Networks · Remittances flows · Economic and Financial Data

1 Introduction

Modelling and understanding longitudinal and temporal data in economics is often challenging due to the complexity and interdependence of various factors. Predicting remittance flows and understanding their dynamics is crucial in the 21st century, marked by massive migration. In many countries, remittances are a significant financial inflow, increasing household income and reducing poverty. During times of political instability, economic downturns, or natural disasters, migrants send more remittances to their home countries, offering an essential economic lifeline [1]. According to the World Bank, remittances to low- and middle-income countries are estimated to have reached $669 billion in 2023, up 3.8%, as advanced economies and Gulf countries continue to help migrants send money home.

The ability to accurately model and predict financial flows is critical in economic and financial analysis. The complexity of remittance flows, influenced by various factors and their interactions, presents a significant challenge. In Sub-Saharan Africa, the impact of remittances on economic activity highlights their crucial role in enhancing recipient economies, especially in sectors with strong financial intermediation [2]. However, the model's complexity and dependence on large amounts of data make it difficult to replicate across different contexts or countries.

Oeking and Choo's analysis of remittances during the COVID-19 pandemic highlights the limitations of relying on historical data and regression models. Predictive challenges arise because historical trends may not fully capture future remittances due to the pandemic's unique global impact and potential structural changes in economies and migration patterns [3].

Utilizing a Multilayer Perceptron (MLP) architecture, the accuracy of the Neural Network model in forecasting remittances for Bangladesh outperforms that of the ARIMA model based on several model selection criteria, as indicated by the study [4]. Using the Box-Jenkins ARIMA Methodology, a declining trend in remittance inflows is identified for the period 2018–2027. However, the model's potential lack of adaptability to sudden economic shifts and assumption-based limitations may not fully capture real-world complexities [5]. Remittances are a significant source of external funding, promoting economic growth by increasing household income and spending.

Panel data regression models may fail to adequately capture the variability in remittance impacts across different economies, oversimplifying complex dynamics [6]. In contrast, Chami found a negative association between remittances and GDP growth, indicating that remittances are not profit-driven and may not be intended to serve as a source of capital for economic development [7].

This paper utilizes the NeuDen model [8], an innovative computational architecture developed to address the limitations of conventional economic modelling methods. The NeuDen model, initially developed for broad economic and financial analysis, is specifically fine-tuned in this study to analyze remittance inflows to Lebanon—a country heavily reliant on remittances amidst an unprecedented economic crisis—a compelling case for this study, and to dissect the intricate patterns of remittances. The NeuDen model combines spatio-temporal learning with fuzzy logic to predict remittance inflows while providing insights into the economic factors that affect them. The bilateral remittance flows from Saudi Arabia to Lebanon offer an important dataset for the model's analysis.

The primary contribution of this work to the field of Artificial Intelligence (AI) is the development and application of the NeuDen AI computational architecture, which integrates evolving Spiking Neural Networks (eSNNs) with evolving Dynamic Neuro-Fuzzy Systems (eDNFS) for predictive and explainable modelling of economic data. This hybrid approach enhances the interpretability and accuracy of models handling complex temporal data, such as remittance flows. The NeuDen model demonstrates superior predictive performance and provides insights into the underlying economic processes, making it a valuable tool for AI researchers and economists.

Following this introduction, Sect. 2 presents a brief review of current issues in economic modelling. Section 3 describes the structure and the functionality of the NeuDen architecture. In Sect. 4, a NeuDen model is developed and applied to a case study focusing on bilateral remittance inflows from Saudi Arabia to Lebanon. Finally, Sect. 5 provides a discussion of the results and suggests future research directions.

2 Current Issues and Techniques for Data Modelling in Economics

Economic models, or road maps of reality, are designed to enhance our understanding of economic processes and their interactions [9]. In the era of big and multidimensional data, there is a need for advanced models that can analyze complex data structures. Economic modelling includes various methods and techniques designed to analyze dynamics and predict the behaviour of economic systems. Some methods used in economic modelling include:

2.1 Regression Analysis in Economic Forecasting

Regression analysis is essential in economic modelling for identifying behavioral relationships between economic variables. Advanced techniques, such as multivariate and nonlinear models, are frequently used in economic forecasting to capture the complex relationships within economic data. These models are vital for understanding the intensity and direction of connections between variables and for forecasting future outcomes.

2.2 Advancements in Time-Series Analysis

A time series consists of observations recorded sequentially at specific intervals [10]. Time series analysis is used in economic models to study the relationships between different economic variables and how they influence each other. It is crucial for understanding trends and patterns in economic data, enabling forecasting future values based on historical observations. Time series models are evaluated based on the accuracy of their forecasts. Adjacent observations in a time series are interdependent, and various approaches are used to analyze this dependence [11]. The NeuDen model builds on traditional time-series analysis by incorporating spatio-temporal learning and neuro-fuzzy systems, capturing both temporal dependencies and the evolving nature of economic data. This approach allows for more accurate predictions and a deeper understanding of the underlying economic processes.

2.3 The Evolution of Neural Network Models in Economics

Artificial neural networks (ANNs) are powerful models that significantly outperform linear regression techniques because they capture nonlinearity in economic data [12]. They can acquire knowledge and learn dependencies between variables based on a sample of observations and then generalize to other samples [13, 14]. Various ANN models have been designed to forecast economic time series data. Data mining techniques in machine learning have been utilized to uncover the recent relevant variables with the most significant predictive ability [15]. A novel approach to predictive and interpretable modelling of multimodal streaming data involves leveraging spiking neural networks (SNNs) to integrate time series data with real-time news updates [16]. The methodology emphasizes the incremental training of the SNN model, enabling efficient processing of multivariate online data. It also enables the use of low-energy neuromorphic hardware for enhanced computational performance. Advanced ANNs demonstrate their potential economic time series modelling [17]. Within dynamic and non-stationary settings, a Spiking Neural Network, tailored for forecasting time series, demonstrated superior performance compared to conventional methods such as Multilayer Perceptron, dynamic Ridge Polynomial, and linear predictor coefficient models, particularly in forecasting financial and economic data [18]. The ability of a complex dynamic systems approach to learn online and update incrementally with time series data demonstrates its potential beyond traditional statistical methods [19]. A general graph neural network framework has been developed to model multivariate time series data and to extract uni-directed relations among variables through a graph learning module. This framework integrates external knowledge, such as variable attributes, and captures spatial and temporal dependencies [20]. Recent advancements in graph convolutional networks have demonstrated their potential to enhance predictive accuracy in dynamic systems, especially when handling spatio-temporal data [21]. By constructing input-output mappings based on human knowledge and data pairs, a hybrid learning procedure is utilized in the Adaptive-Network-based Fuzzy Inference System (ANFIS) to model nonlinear functions, identify nonlinear components, and predict chaotic time series [22]. A unified soft computing framework has been developed, including rule extraction, refinement, and fuzzy reasoning and control. This framework highlights the potential for rule generation from fuzzy knowledge-based networks, emphasizing its applicability to real-world scenarios [23]. Spatio-temporal learning models are essential for analyzing complex, dynamic data, with metrics like precision and recall providing insights into model performance in capturing temporal dependencies [24]. A unique method (PAMeT) is intended to generate predictive associative memories, learn temporal associations over a complete dataset of time series variables, and efficiently capture complex data dynamics. It permits more accurate forecasting of future values with fewer available independent features [25].

2.4 NeuDen: A New Paradigm in Economic Modelling

Adam Smith introduced the term "*invisible hand*" to describe the complexity of economic phenomena. In his work, "*The Wealth of Nations*," Smith emphasized the economy's inherent ability to govern itself. Today, economists use various models to better understand how this invisible hand influences real-world dynamics [26]. This paper employs an advanced method in economic modelling, the NeuDen, which significantly enhances our ability to capture and analyze the interactions explained by Adam Smith's concept of the invisible hand.

NeuDen is a generic computational architecture that provides an advanced toolset for understanding complex patterns in time series data. With its strong analytical capabilities, we can better interpret and apply Smith's insights to enhance real-world economic applications. Specifically, we employ NeuDen's computational architecture [8] for predictive modelling in economics as it offers a novel approach to handling complex economic data, making it perfect for our remittance prediction research.

3 The Structure and Functionality of the NeuDen Model

The NeuDen model represents a sophisticated evolution within the domain of Evolving Connectionist Systems (ECOS) [27], standing at the intersection of evolving spiking neural networks (eSNNs) and evolving dynamic neuro-fuzzy systems (eDNFS).

Since their foundation by Professor Nikola Kasabov, Evolving Connectionist Systems (ECOS) [28] have transformed how we approach learning in dynamic settings. As exemplified by pioneering models like NeuCube [29] and DENFIS [30], ECOS is commended for its ability to adapt and learn in real-time as new data arrives. This makes ECOS ideal for handling complex non-stationary data, as it can capture detailed patterns and relationships that static models may not [16, 31–36]. NeuCube, an ECOS example, has shown remarkable efficiency in spatio-temporal data learning, while DENFIS has proven useful in dynamic time-series prediction. Their achievement demonstrates the continued relevance of ECOS principles in modern computational intelligence.

NeuDen builds on this tradition by leveraging the ECOS framework's core principle of continual learning and adaptation. Designed to process and analyze complex, high-dimensional datasets, NeuDen represents a hybrid spatio-temporal neuro-fuzzy approach. This integration positions the NeuDen model as a powerful tool for analyzing and interpreting data with an evolving nature, such as financial and economic datasets, particularly in the context of remittance flows.

The NeuDen model is divided into three principal components (Fig. 1), each embodying the core principles of ECOS by facilitating continuous learning, adaptation, and evolution in response to new data inputs and processing dynamics of temporal remittance flows.

o *Spatio-Temporal Learning Module:* This component utilizes evolving spiking neural networks (eSNNs) techniques to convert raw time-series data into binary spikes. These spikes then undergo both unsupervised and supervised learning processes for temporal and spatial analysis.
o *Vector Transformation Module:* This module bridges the gap between spatio-temporal and neuro-fuzzy modelling by converting spike-encoded data into a structured vector format, thus facilitating the transition from temporal encoding to fuzzy inference.
o *Neuro-Fuzzy Inference Module:* This component employs a dynamic evolving neuro-fuzzy inference system to learn and analyze the transformed vector data, enabling the model to make predictions based on learned temporal patterns and fuzzy logic rules.

The learning methodology of the NeuDen model is characterized by its evolving nature, incremental learning capability, and dynamic adaptability to incoming data streams. By integrating neuro-fuzzy inference with spatio-temporal learning, the NeuDen model adopts a hybrid approach that harnesses the strengths of both methodologies to enhance its data analysis capabilities.

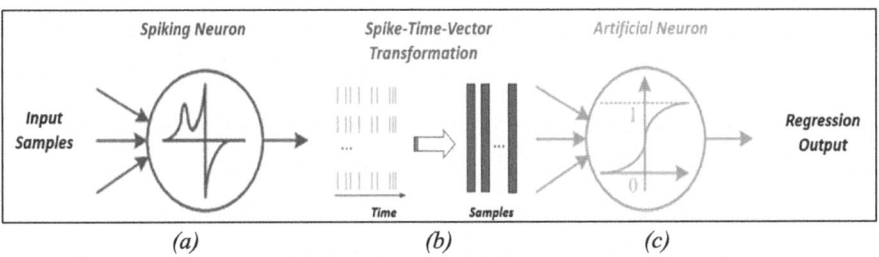

Fig. 1. The NeuDen Principal Diagram. (a) Spatio-Temporal Learning Module, (b) Vector Transformation Module, (c) Neuro-Fuzzy Inference Module (from [20]).

o *Spatio-Temporal Learning:* At the core of the NeuDen model's learning methodology is the spatio-temporal learning process, which utilizes the principles of spike-time-dependent plasticity (STDP) within the unsupervised learning mechanism to adjust synaptic weights based on the timing of incoming spikes. Furthermore, the dynamic evolving Spiking Neural Network (deSNN) supervised learning enhances STDP learning by incorporating dynamic synapses through Rank-Order and Spike Driven Synaptic Plasticity learning rules, thus providing a robust foundation for subsequent fuzzy inference.

o *Neuro-Fuzzy Inference Learning:* Building upon the temporal patterns encoded by the SNNs, the NeuDen model employs a neuro-fuzzy inference learning process to analyze and interpret data within a fuzzy logic framework. This involves formulating fuzzy rules and their adjustments based on incoming data, allowing the model to refine its predictions over time.

4 Developing a NeuDen Model for Predictive Modelling of Bilateral Remittances Inflows

This paper introduces an adaptation of the NeuDen model, focusing solely on the analysis of bilateral remittances inflows. Utilizing a novel 'NeuDen' approach within a spatio-temporal neuro-fuzzy framework, we analyze the complex dynamics of remittances sent from Saudi Arabia to Lebanon. The model leverages evolving spiking neural networks (eSNNs) to encode temporal features and neuro-fuzzy systems for interpretability and prediction accuracy. Our findings not only underline the model's effectiveness in capturing the nuanced patterns of remittance inflows but also offer insights into the factors influencing these financial flows. By focusing on remittances, this study contributes to the understanding of economic resilience and the strategic importance of diaspora contributions in the face of economic challenges.

4.1 Data Description

Remittances, a vital source of income in developing countries, have been increasingly recognized for their role in economic stability and growth. Remittances denote the income received by households from foreign economies, primarily stemming from individuals' temporary or permanent relocation to those economies. Remittances are small, regular amounts sent by an individual working in another country, represented here as a host economy, to his family or relatives in his country of origin. They are a form of familial aid that is used in many countries.

Figure 2 shows high levels of remittance inflows to Lebanon from Saudi Arabia during the period 2014–2017, with a peak in 2017 at $1.6 billion.

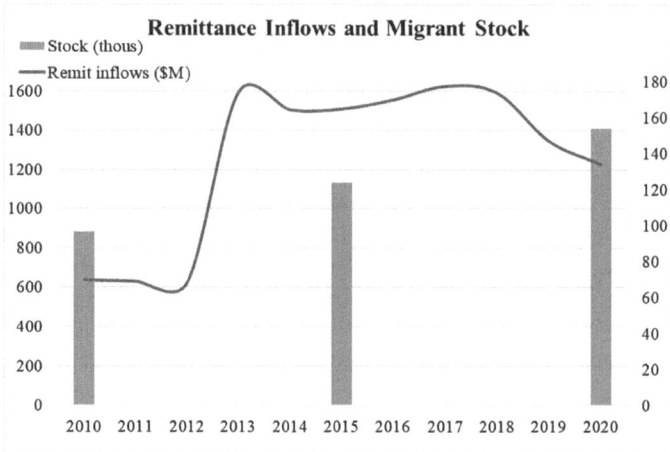

Fig. 2. Remittances inflows from SA to LB and Migrant Population of LB in SA. Source: World Bank and UNDESA.

The migrant population has grown substantially in recent years, suggesting that an increasing number of migrants may lead to larger remittance inflows. However, remittances dropped between 2019 and 2020, coinciding with significant economic challenges in Lebanon, including a financial crisis and the COVID-19 pandemic, which likely contributed to this decline. By 2020, there were 154 thousand migrants in the country, indicating that other economic factors also influence remittances beyond just the number of migrants.

4.2 Dataset and Pre-Processing

The remittance dataset comprises 120 temporal features spanning from January 2011 to December 2020. The spatial features include remittances inflows to Lebanon (LB) from Saudi Arabia (SA) extracted, in millions of U.S. Dollars, from the Global Knowledge Partnership on Migration and Development (KNOMAD) implemented by the World Bank [37]. These bilateral remittance inflows are estimated by KNOMAD on an annual basis. The stock of Lebanese migrants in SA is extracted from the UNDESA [38]. The stock is estimated on a five-year period basis, and the conversion to monthly constant distribution is done for the model purposes. The GDP and current prices in billions of U.S. dollars for SA and LB are extracted on an annual basis from the International Monetary Fund (IMF) portal [39]. Generating monthly GDP positions and remittance inflows is done for modelling purposes using a proportional allocation method. The official CPI data for Lebanon is extracted on a monthly basis from Lebanon's Central Administration of Statistics [40] (Table 1).

In this case study, however, due to the high volatility and nonlinearity of remittance flows, the original data have been normalized using the simple min-max scaling technique. Then, a transformation of 108 window samples of size 12 temporal features is performed. Thus, we aim to analyze and predict remittance inflows based on their historical values and other chosen variables.

Table 1. The Economic Variables as Input Features.

Spatial feature	Description
rem	Remittance inflows SA to LB (US$ millions)
stock	Stock of Lebanese migrants in SA (thousands)
gdp.0	GDP of Country of Origin [LB] (US$ billions)
gdp.h	GDP of Host Country [SA] (US$ billions)
Cpi	Consumer Price Index of LB (units)

4.3 Spatio-Temporal Modelling

The spatio-temporal bloc of the NeuDen model uses evolving Spiking Neural Networks (eSNNs) with specific parameters. A Threshold of 0.5 is employed to encode the generated samples into trains of spikes that are fed to this bloc. A split of 50/50 is applied for training and incremental learning and testing procedures. A total of 1000 neurons are set to train the 3D cube with a minimum connectivity of 2.5 radius. In this bloc, the eSNN is first trained incrementally in an unsupervised mode using the STDP learning rate set to 0.01, leaking rate = 0.002, firing threshold = 0.5, and refractory period = 6. The STDP learning rule considers the time of spiking between two connected neurons so that consecutive changes in data from one time to another across all input variables are learned.

After applying the unsupervised learning, the NeuDen model adjusts the synaptic weights based on the time difference between the pre-synaptic and the post-synaptic spikes. This process ensures that the model accurately captures the temporal dependencies within the data. Figure 3 provides a two-dimensional visualization of these adjusted connections (weights) between activated neurons. The 0.08 weight represents a threshold for strong connections, indicating a significant influence on remittance inflows. It is clear that remittance inflows (REM), migrant stock (STOCK), and GDP of the host country are highly active and influence the inflows of remittances, which is the feature under study. The GDP of the original country and the corresponding CPI are less influential. It is worth noting that in this case study, the model is highly active, where almost all neurons are super active, as indicated by the very bright colour of the activated neurons.

In the supervised learning phase, the connections between neurons are further refined based on the Rank-Order and Spike-Driven Synaptic Plasticity rules, identified with the following parameters: modulation factor (0.8) and drift parameter (0.005). These learning rules enhance the model's ability to predict remittance inflows with a high degree of accuracy, as demonstrated by the 86% accuracy rate depicted in Fig. 4.

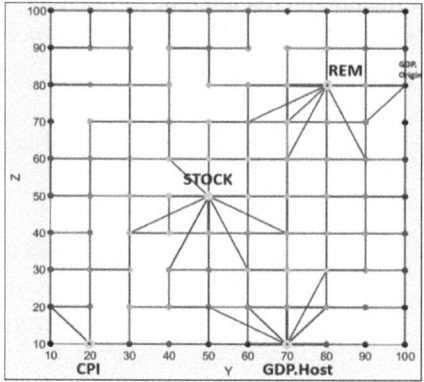

Fig. 3. Spatio-temporal connections above 0.08 weight in the 2D cross-section.

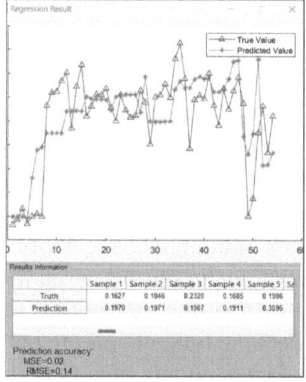

Fig. 4. Regression after spatio-temporal learning.

4.4 Neural-Fuzzy Inference Modelling

Following the spatio-temporal learning phase, the spike rates are extracted in a vector-based format, with a dimension of 108 samples for the five features. In this case study, the spike trains are subsequently fed into the neuro-fuzzy inference system without applying any normalization to the input vector. The absence of normalization in this phase is intentional, allowing the model to handle raw, nonlinear data without imposing any external constraints that could potentially obscure the intrinsic relationships within the data and bias the results. The fuzzy inference module then uses these vectors to generate fuzzy rules that help interpret the temporal patterns encoded by the spiking neural networks. This approach leverages the particularity of the NeuDen model when the goal is to model complex, nonlinear interactions, as it avoids introducing biases that could distort the model's predictions.

However, some cleaning to the vector data is applied, where zero samples are cut off for the purposes of the neuro-fuzzy modelling. Figure 5 plots the spike vector of the REM variable for input to fuzzy learning. The whole dataset is being trained and incrementally tested with a threshold set at 0.9, identified by three fuzzy membership functions to estimate the output of the sampled data and two learning iterations.

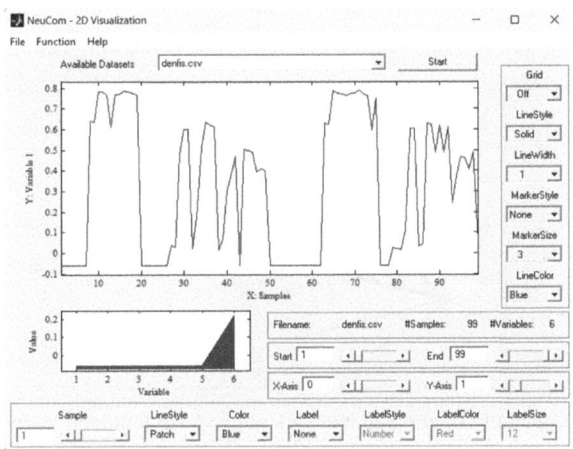

Fig. 5. 2D visualization of the input signal extracted from spatio-temporal learnings.

4.5 Experimental Results and Analysis

The Feature Interaction Network, after the spatio-temporal learning phase, shows the strong influence between remittance inflows (REM) and migrant stock (STOCK) variables (Fig. 6). The analysis reveals that REM is slightly affected by the GDP of the migrant's country of origin. Conversely, a high GDP in the host country attracts more migrants, allowing them to send more money to their families.

The spike vectors, which are derived from the spatio-temporal learning phase, are further analyzed using a fuzzy logic system that enables the model to generate explainable rules to best describe the interaction between different economic variables, such as remittance inflows, GDP, and migrant stock. This explainability is a key feature of the NeuDen model, allowing for both accurate predictions and a deeper understanding of the underlying economic processes.

Figure 7 demonstrates the regression curve of the NeuDen model after the fuzzy logic training, showcasing a marked enhancement in precision. The model's accuracy improved by a multiple factor of 3.4 compared to the spiking neural network (SNN) output represented in Fig. 4, recording a significant accuracy level of 95.8% when employing the Leave-one-out cross-validation method.

4.6 Comparison Between NeuDen and Other Machine Learning Methods

Additional analyses on the normalized dataset using various regression methods have been performed to further confirm the NeuDen model's superiority. The models' accuracy is evaluated using RMSE, where a lower RMSE indicates better predictive accuracy and model robustness.

Table 2 highlights the superior predictive power of the NeuDen model with an RMSE of 0.04 using a 50/50 data split for training and incremental testing, markedly outperforming other methods such as Multiple Regression, Logarithmic Regression, KNN, DENFIS, Multilayer Perceptron, and NeuCube.

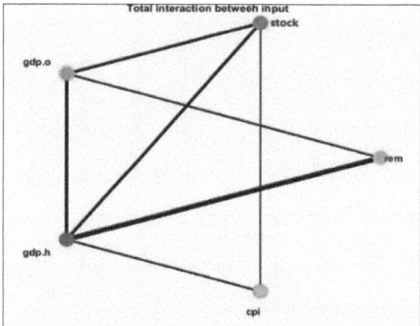

Fig. 6. Feature Interaction Network after spatio-temporal learning.

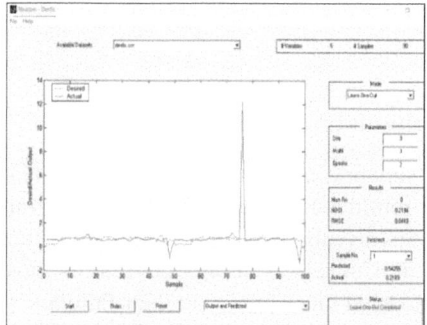

Fig. 7. Regression outcomes post fuzzy logic learning and incremental testing phase.

This advantage highlights the model's ability to capture complex patterns and interactions in the data, making it an effective tool for economic analysis and forecasting.

NeuDen's learning process is performed on binary representations that capture fluctuations in input features over consecutive time points. It is pivotal to note that the model's predictive accuracy, as revealed in the regression output after the neuro-fuzzing training, is based on the variances encoded within the input spike vector. These spikes capture the change in the original dataset rather than the raw data itself.

Consequently, the regression analysis indicates that more substantial changes in both the migrant stock and the GDP of the host country have a correlative improvement effect on remittance volumes. Simultaneously, a downturn in the GDP of the origin country, typically the remittance-receiving country, unexpectedly correlates with a strengthening, or at least a stabilization, of remittance inflows (Eq. 1).

The coefficients in the equation represent the influence of each variable on remittance inflows, with positive or negative values indicating the direction of this influence. For instance, for every one-unit increase in the GDP of the host country (GDP_h), remittance inflows increase by 3.15 units. This shows that stronger economic performance in the host country positively affects the amount of money migrants can send home. Similarly, the stock of migrants has a positive effect, and the previous period's remittances positively influence current remittances. However, the regression formula shows that a decrease in the GDP of the origin country (GDP_o) would suggest a compelling economic behavior where, particularly in Lebanon's case, remittances tend to either maintain or increase as the domestic GDP contracts, which reflect a protective financial response among the diaspora as economic conditions at home worsen.

$$REM = 1.51 + 0.16 * REM_{-1} + 0.03 * Stock - 3.34 * GDP.o + 3.15 * GDP.h - 0.07 * CPI \tag{1}$$

Table 2. The Level of Accuracy as measured by the RMSE.

Method	Split	RMSE
Multiple Regression	Whole dataset	0.13
Logarithmic Regression	Whole dataset	0.13
KNN for prediction	Whole dataset	0.10
DENFIS	50/50	0.28
Multilayer Perceptron	50/50	0.17
NeuCube	50/50	0.14
NeuDen	50/50	0.04

5 Conclusion

The paper introduces the NeuDen model, a novel computational architecture that marks a new paradigm in adaptive and evolving systems, specifically designed for predictive modelling and the explanation of dynamic time series data in economics. The NeuDen model leverages the advanced capabilities of evolving Spiking Neural Networks (eSNNs) and evolving Dynamic Neuro-Fuzzy Systems (eDNFS), enabling it to manage and analyze complex data patterns, visualize intricate data connectivity, and produce accurate regression outputs through fuzzy inference methods.

The NeuDen model exemplifies the never-ending cycle of innovation and adaptability inherent in evolving intelligent systems. Through the case study on remittance flows, this research has demonstrated the practical advantages of the NeuDen approach, showing its potential to address a wide range of challenges in economics and finance.

Moreover, by bridging the gap between interpretability and predictive accuracy, NeuDen contributes significantly to the AI field, particularly in economic forecasting and the understanding of dynamic financial systems, as well as modelling complex temporal economic data.

Beyond economics, the NeuDen approach can potentially be useful for solving problems in other areas, such as cybersecurity, by analyzing network traffic anomalies and predicting cyberattacks. Another promising area is neuroscience and brain-machine interfaces [41, 42]. This versatility underscores the broader applicability of the NeuDen model, positioning it as a valuable tool for tackling complex problems across various domains. In addition to new applications, the NeuDen method can also stimulate other methods of computational intelligence, such as clustering of temporal data [43], spatio-temporal associative memories [44], and life-long learning [45].

Acknowledgements. We gratefully acknowledge that this research is conducted and funded in relation to the execution of a scientific research project № КП-06-Н35/12, "An Innovative Approach in Developing an Intelligent Information System for Detection and Prevention of Financial and Customs Fraud," under the contract with National Science Fund in Bulgaria. We also acknowledge that the research and the preparation of the paper are sponsored by Knowledge-Engineering Ltd. (www.knowledgeengineeing.ai) [46]. Additionally, we would like to note that the NeuCube [47] and NeuCom [48] software are both available for free for research and innovation on the KEDRI website.

Disclosure of Interests. The authors declare no competing interests relevant to the content of this article. Furthermore, the authors did not receive any funding, grants, or other support from any organization for the conduct and submission of this study.

Disclaimer. The concepts, ideas, and content in this work are the intellectual property of the authors and are related solely to their contributions and perspectives. Our affiliated institutions explicitly disclaim any responsibility for the interpretations or applications of the information presented herein. The data utilized in this research is accessible through the corresponding citations provided within the text.

References

1. Ratha, D.: Remittances: a lifeline for development. Int. Monetary Fund Finan. Dev. **42**, 4 (2005)
2. Perez-Saiz, H., Dridi, J., Gursoy, T., Bari, M.: The impact of remittances on economic activity: the importance of sectoral linkages. Int. Monetary Fund **175** (2019). https://doi.org/10.5089/9781498324489.001
3. Oeking, A., Choo, C.Y.: Coming home: are remittances in the ASEAN+3 another victim of the pandemic? (2020). https://amro-asia.org/coming-home-are-remittances-in-the-asean3-another-victim-of-the-pandemic/

4. Hossain, M., Abdulla, F., Majumder, A.: Comparing the forecasting performance of ARIMA and neural network model by using the remittances of Bangladesh. Jahangirnagar Univ. J. Stat. Stud. **34**, 1–12 (2017)
5. Nyoni, T.: Modelling and Forecasting Remittances in Bangladesh Using The Box-Jenkins ARIMA Methodology (2019)
6. Meyer, D., Shera, A.: The impact of Remittances on Economic growth: an econometric model. Economi A.18 (2016). https://doi.org/10.1016/j.econ.2016.06.001
7. Chami, R., Fullenkamp, C., Jahjah, S.: Are immigrant remittance flows a source of capital for development? IMF Staff Papers **52**(1), 55–81 (2005). http://www.jstor.org/stable/30035948
8. Abouhassan, I., Kasabov, N.: NeuDen: a framework for the integration of neuromorphic evolving spiking neural networks with dynamic evolving neuro-fuzzy systems for predictive and explainable modelling of streaming data. Evolving Syst. **16**, 3 (2025). https://doi.org/10.1007/s12530-024-09630-4
9. Ouliaris, S.: What are economic models. Int. Monetary Fund Finan. Dev. **48**(2) (2011)
10. Brockwell, P.J., Davis, R.A.: Introduction to Time Series and Forecasting. 2nd ed. Springer, New York (2002). Box, G., Jenkins, G., Reinsel, G., Ljung, G.: Time series analysis: forecasting and control. J. Time Ser. Anal. 5th ed., Wiley and Sons Inc., Hoboken, New Jersey, p. 712 (2015). https://doi.org/10.1111/jtsa.12194
11. Aminian, F., Suarez, E.D., Aminian, M., et al.: Forecasting economic data with neural networks. Comput. Econ. **28**, 71–88 (2006). https://doi.org/10.1007/s10614-006-9041-7
12. Herbrich, R., Keilbach, M., Graepel, T., Bollmann-Sdorra, P., Obermayer, K.: Neural networks in economics. In: Brenner, T. (eds.) Computational Techniques for Modelling Learning in Economics. Advances in Computational Economics, vol. 11. Springer, Boston, MA (1999). https://doi.org/10.1007/978-1-4615-5029-7_7
13. Li, Y., Ma, W.: Applications of artificial neural networks in financial economics: a survey. In: International Symposium on Computational Intelligence and Design, Hangzhou, China, pp. 211–214 (2010). https://doi.org/10.1109/ISCID.2010.70
14. Kaastra, I., Boyd, M.: Designing a neural network for forecasting financial and economic time series. Neurocomputing **10**(3), 215–236 (1996). https://doi.org/10.1016/0925-2312(95)00039-9
15. Thawornwong, S., Enke, D.: The adaptive selection of financial and economic variables for use with artificial neural networks. Neurocomputing **56**, 205–232 (2004). https://doi.org/10.1016/j.neucom.2003.05.001
16. Abouhassan, I., Kasabov, N., Jagtap, V., et al.: Spiking neural networks for predictive and explainable modelling of multimodal streaming data with a case study on financial time series and online news. Sci. Rep. **13**(18367), 2023 (2023). https://doi.org/10.1038/s41598-023-42605-0
17. Falat, F., Pancikova, L.: Quantitative modelling in economics with advanced artificial neural networks. Proc. Econ. Finan. **34**, 194–201 (2015). https://doi.org/10.1016/S2212-5671(15)01619-6
18. Reid, D., Hussain, A.J., Tawfik, H.: Financial time series prediction using spiking neural networks. PLoS ONE **9**(8), e103656 (2014). https://doi.org/10.1371/journal.pone.0103656
19. Widiputra, H., Pears, R., Serguieva, A., Kasabov, N.: Dynamic interaction networks in modelling and predicting the behaviour of multiple interactive stock markets. Intell. Syst. Account. Finan. Manag. **16**, 189–205 (2008). https://doi.org/10.1002/isaf.300
20. Wu, Z., Pan, S., Long, G., Jiang, J., Chang, X., Zhang, C.: Connecting the dots: multivariate time series forecasting with graph neural networks. Assoc. Comput. Mach. (2020). https://doi.org/10.1145/3394486.3403118.ISBN 9781450379984
21. Zhang, S., Tong, H., Xu, J., et al.: Graph convolutional networks: a comprehensive review. Comput. Soc. Netw. **6**, 11 (2019). https://doi.org/10.1186/s40649-019-0069-y

22. Jang, J.: ANFIS: adaptive-network-based fuzzy inference system. IEEE Trans. Syst. Man Cybern. **23**(3), 665–685 (1993). https://doi.org/10.1109/21.256541
23. Mitra, S., Hayashi, Y.: Neuro-fuzzy rule generation: survey in soft computing framework. IEEE Trans. Neural Netw. **11**(3), 748–768 (2000)
24. Powers, D.: Evaluation: From Precision, Recall and F-Factor to ROC, Informedness, Markedness & Correlation. Mach. Learn. Technol 2 (2008)
25. AbouHassan, I., Kasabov, N., et al.: ePAMeT: evolving Predictive Associative Memories for Time series. Evolving Syst. **16**, 6 (2025). https://doi.org/10.1007/s12530-024-09628-y
26. Smith, A.: An Inquiry into the Nature and Causes of the Wealth of Nations, McMaster University Archive for the History of Economic Thought (1776)
27. Kasabov, N.: Time-Space, Spiking Neural Networks and Brain-Inspired Artificial Intelligence. Springer Series on Bio - and Neurosystems (SSBN, volume 7) (2019). https://doi.org/10.1007/978-3-662-57715-8
28. Kasabov, N.: The ECOS framework and the ECO learning method for evolving connectionist systems. J. Adv. Comput. Intell. **2**(6), 195–202 (1998)
29. Kasabov, N.: NeuCube: a spiking neural network architecture for mapping, learning and understanding of spatio-temporal brain data. Neural Netw. **52**, 62–76 (2014). https://doi.org/10.1016/j.neunet.2014.01.006. Elsevier
30. Kasabov, N., Song, Q.: DENFIS: dynamic evolving neural-fuzzy inference system and its application for time-series prediction. IEEE Trans. Fuzzy Syst. **10**, 144–154 (2002). https://doi.org/10.1109/91.995117
31. Kasabov, N.: Evolving fuzzy neural networks-algorithms, applications and biological motivation. In: Methodologies for the Conception, Design and Application of Soft Computing, pp. 271–274 (1998)
32. Kasabov, N.: Adaptation and interaction in dynamical systems: modelling and rule discovery through evolving connectionist systems. Appl. Soft Comput. **6**, 307–322 (2006). https://doi.org/10.1016/j.asoc.2005.01.006
33. Kasabov, N., Dhoble, K., Nuntalid, N., Indiveri, G.: Dynamic evolving spiking neural networks for online spatio - and spectro-temporal pattern recognition. Neural Netw. **41**, 188–201 (2013). https://doi.org/10.1016/j.neunet.2012.11.014
34. Schliebs, S., Kasabov, N.: Computational modelling with spiking neural networks. Springer Handbook of Bio-/Neuro Informatics, pp. 625–646 (2014)
35. Abouhassan, I., Kasabov, N., Popov, G., Trifonov, R.: Why use evolving neuro-fuzzy and spiking neural networks for incremental and explainable learning of time series? a case study on predictive modelling of trade imports and outlier detection. In: IEEE 11th International Conference on Intelligent Systems (IS), Warsaw, Poland, pp. 1–7 (2022). https://doi.org/10.1109/IS57118.2022.10019673
36. Takagi, T., Sugeno, M.: Fuzzy identification of systems and its applications to modelling and control. IEEE Trans. Syst. Man Cybern. **1**, 116–132 (1985)
37. World Bank - KNOWMAD Homepage. https://www.knomad.org/data/remittances. Accessed 6 April 2024
38. United Nations Department of Economic and Social Affairs, Population Division (2020). International Migrant Stock (2020). https://www.un.org/development/desa/pd/content/international-migrant-stock. Accessed 6 April 2024
39. International Financial Statistics Homepage. https://data.imf.org/regular.aspx?key=61545850. Accessed 6 April 2024
40. Central Administration of Statistics Homepage. http://www.cas.gov.lb/index.php/national-accounts-en. Accessed 6 April 2024
41. Kelsen, B.A., Sumich, A., Kasabov, N., Liang, S.H.Y., Wang, G.Y.: What has social neuroscience learned from hyper-scanning studies of spoken communication? a systematic review.

Neurosci. Biobehav. Rev. **132**, 1249–1262 (2022). https://doi.org/10.1016/j.neubiorev.2020.09.008. ISSN 0149-7634

42. Kumarasinghe, K., Kasabov, N., Taylor, D.: Brain-inspired spiking neural networks for decoding and understanding muscle activity and kinematics from electroencephalography signals during hand movements. Sci. Rep. **11**(1), 2486 (2021). https://doi.org/10.1038/s41598-021-81805-4
43. Tu, E., Cao, L., Yang, J., Kasabov, N.: A novel graph-based k-means for nonlinear manifold clustering and representative selection. Neurocomputing **143**, 109–122 (2014). https://doi.org/10.1016/j.neucom.2014.05.067
44. Kasabov, N.K.: STAM-SNN: spatio-temporal associative memories in brain-inspired spiking neural networks: concepts and perspectives. In: Kovács, L., Heidegger, T., Szakál, A. (eds.) Recent Advances in Intelligent Engineering. Topics in Intelligent Engineering and Informatics, vol. 18. Springer (2024). https://doi.org/10.1007/978-3-031-58257-8_1
45. Kasabov, N.K.: Life-long learning and evolving associative memories in brain-inspired spiking neural networks. MOJ App. Bio. Biomech. **8**(1), 56–57 (2024). https://doi.org/10.15406/mojabb.2024.08.00208
46. Knowledge Engineering Homepage. http://www.knowledgeengineering.ai
47. NeuCube Homepage. http://www.kedri.aut.ac.nz/neucube
48. NeuCom Homepage. http://www.theneucom.com

Deep Learning for Multi-class Diagnosis of Thyroid Disorders Using Selective Features

Filipa Santana[1(✉)], José Brito[1,2], and Petia Georgieva[3]

[1] University of Aveiro, Aveiro, Portugal
vfssantana@ua.pt
[2] INESC TEC, Centre for Power and Energy Systems, Porto, Portugal
jose.brito@inesctec.pt
[3] Instituto de Telecomunicações, Aveiro/IEETA/Department of Electronics, Telecommunications and Informatics, University of Aveiro, Aveiro, Portugal
petia@ua.pt

Abstract. Data-based approach for diagnosis of thyroid disorders is still at its early stage. Most of the research outcomes deal with binary classification of the disorders, i.e. presence or not of some pathology (cancer, hyperthyroidism, hypothyroidism, etc.). In this paper we explore deep learning (DL) models to improve the multi-class diagnosis of thyroid disorders, namely hypothyroid, hyperthyroid and no pathology thyroid. The proposed DL models, including DNN, CNN, LSTM, and a hybrid CNN-LSTM architecture, are inspired by state-of-the-art work and demonstrate superior performance, largely due to careful feature selection and the application of SMOTE for class balancing prior to model training. Our experiments show that the CNN-LSTM model achieved the highest overall accuracy of 99%, with precision, recall, and F1-scores all exceeding 92% across the three classes. The use of SMOTE for class balancing improved most of the model's performance. These results indicate that the proposed DL models not only effectively distinguish between different thyroid conditions but also hold promise for practical implementation in clinical settings, potentially supporting healthcare professionals in more accurate and efficient diagnosis.

Keywords: Thyroid disorders · hypothyroid · hyperthyroid · Feature selection · Deep Neural Networks (DNN) · Convolutional Neural Networks (CNN) · Long Short-Term Memory (LSTM)

1 Introduction

The thyroid gland (Fig. 1a) is a crucial component of the endocrine system, and responsible for producing hormones that regulate several body functions, [1]. The thyroid gland produces two primary hormones, thyroxine (T4), also called T4 because it contains four iodine atoms, and triiodothyronine (T3), which results from the removal of one iodine atom. The hormones T3 and T4 regulate metabolism, energy production and body temperature, and are crucial for the proper functioning of the body organs and systems. Any imbalance in thyroid hormone levels can lead to health problems, affecting metabolism, mood, energy levels and even heart rate.

The amount of T4 produced by the thyroid gland is controlled by another hormone produced in the pituitary gland called thyroid stimulating hormone (TSH). The interaction between the thyroid gland and the pituitary gland is crucial in regulating thyroid hormone production. If the pituitary gland registers a very low T4 value, it produces more TSH to stimulate the thyroid gland to produce more T4. When T4 in the bloodstream exceeds a certain level, the pituitary gland's production of TSH stops [1]. Figure 1b illustrates this cooperation work diagram.

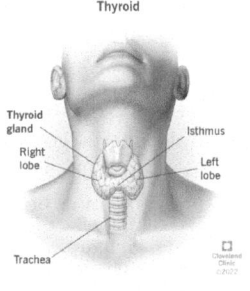

(a) Thyroid gland

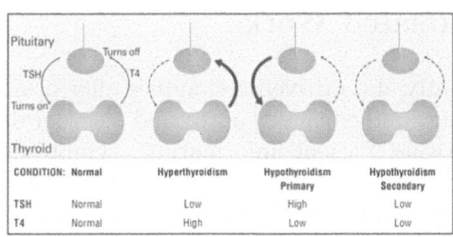

(b) Pituitary and Thyroid interaction schema [1]

Fig. 1. Pituitary and Thyroid

Irregularities in the thyroid gland can lead to different abnormalities, and two of the most common are hyperthyroidism and hypothyroidism [2]. According to [1], Hypothyroidism occurs when the thyroid gland produces insufficient hormones, leading to symptoms such as fatigue, weight gain, and sensitivity to cold. On the other hand, hyperthyroidism results from an overactive thyroid gland, causing symptoms like weight loss, rapid heartbeat, and anxiety. Both conditions can significantly impact an individual's quality of life if left untreated.

Diagnosing thyroid disorders typically involves a combination of medical history assessment, physical examination, and laboratory tests. Blood tests measuring thyroid hormone levels (T3, T4) and TSH are commonly used to assess thyroid function. Changes in TSH can serve as an early warning system and T4 and T3 tests can help doctors to identify the severity and type of disorder. Imaging tests such as ultrasound may also be performed to evaluate the thyroid gland's structure and detect any physical abnormalities.

Diseases of the thyroid gland are among the most prevalent endocrine diseases in the world, second only to diabetes, according to the World Health Organisation. Hyperthyroidism and hypothyroidism affect around 2 per cent and 1 per cent of individuals respectively, and men have around a tenth of the prevalence of women [3].

However, diagnosing thyroid disorders is a challenging task due to overlapping symptoms with other medical conditions and subtle changes in hormone levels. This complexity underscores the need for accurate and timely diagnosis to initiate appropriate treatment and prevent potential complications, [4].

It is in this context that Machine Learning (ML) offers a promising approach to improve accuracy and diagnostic efficiency, and previous research has shown that ML models can significantly improve diagnostic processes (see [2,3,5–9]). These studies explore several ML techniques for thyroid disease classification. By leveraging comprehensive datasets and advanced analytics, we aim to enhance diagnostic accuracy, aiding healthcare professionals and ultimately improving patient care, and enable earlier diagnoses, reducing wait times and optimizing resource allocation.

2 Related Work

Currently, data driven techniques offer promising alternatives for improving the accuracy and efficiency of thyroid disorder diagnosis. By analyzing large datasets comprising patient information, symptoms, and laboratory results, machine learning models can identify patterns and relationships that may not be immediately apparent to human observers.

However, the related work is predominantly focused on binary classification problems where the subjects are classified into thyroid patients or healthy subjects. Multiclass sickness diagnosis is not well studied with only a few results published (see [2,5–7]).

Further to that, the emphasis is usually placed on the optimization of machine or deep learning (ML/DL) models with the feature selection stage being less explored or completely ignored for the thyroid disease problem.

Early predictions only for hypothyroidism are proposed in [8], after feature selection step and based on Support Vector Machine (SVM), Decision Trees (DT), Random Forest (RF), Logistic Regression (LR) and Naive Bayes (NB) classifiers.

In order to compensate this gap in [2] a number of ML/DL models are combined with different feature selection approaches to deal with the multi-class thyroid disease identification problem.

In [6] and [7] several classifiers are compared to classify Thyroid disease into normal, Hypothyroidism, or hyperthyroidism categories, but they didn't use deep learning techniques, they use DT, SVM, NB, and LR.

Jha, Bhattacharjee, and Mustafi [5] applied input dimension reduction techniques and data augmentation in order to generate sufficient data for their deep neural network (DNN) model. The importance rank of the features is defined based on DT models. K-Nearest Neighbour (KNN) and DNN classifiers are trained after the feature reduction.

In the present paper we compare a number of DL models to distinguish between hypothyroidism, hyperthyroidism and normal thyroid states. The models were trained with a large dataset composed of various thyroid related datasets. After a proper feature selection step the model diagnostic performance was enhanced compared to the state of the art works.

3 Dataset

The dataset used in this work is a combination of six datasets, all with information collected at the Garvan Institute in Sydney, Australia, and obtained from the UCI Machine Learning platform [10]. The original data consist of 19373 patient records, 26 features and the class variable. However, not all features are present in all constitution datasets, therefore we have selected the 18 features present in all datasets, summarised in Table 1. After a pre-processing step, where the null values and the outliers were eliminated, we have ended up with 11066 patient records.

Table 1. Feature description

Feature	Description	Data type
age	age of the patient	(int)
sex	sex patient identifies	(str)
on thyroxine	patient is on thyroxine	(bool)
query on thyroxine	query on thyroxine	(bool)
on antithyroid medication	patient is on antithyroid medication	(bool)
sick	if patient is sick	(bool)
pregnant	if patient is pregnant	(bool)
thyroid surgery	if patient has undergone surgery	(bool)
query hypothyroid	believe it has hypothyroid	(bool)
query hyperthyroid	believe it has hyperthyroid	(bool)
lithium	if patient is taking lithium	(bool)
goitre	if patient has goitre	(bool)
tumor	if patient has tumor	(bool)
TSH	TSH level in blood	(float)
T3	T3 level in blood	(float)
TT4	TT4 level in blood	(float)
T4U	T4U level in blood	(float)
FTI	FTI level in blood	(float)
Target	Medical diagnosis (classification)	(str)

In Fig. 2a and Fig. 2b are illustrated the age distribution of the patients diagnosed with hypothyroid and hyperthyroid, respectively. It is evident that these pathologies are diagnosed with higher probability between 55 and 62 years old. According to the literature this type of disorders are more common in women then in men, and the dataset corroborates with these conclusions as shown in Fig. 3a.

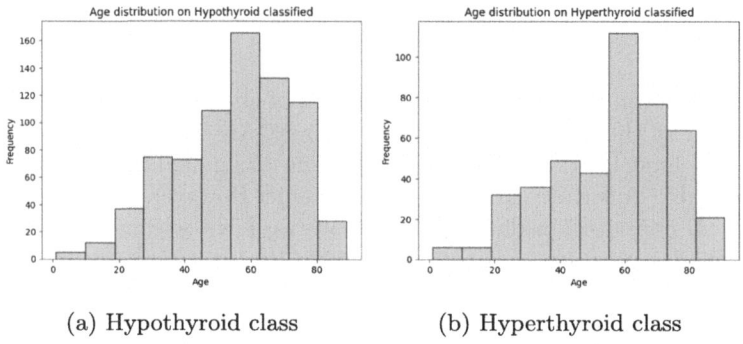

(a) Hypothyroid class (b) Hyperthyroid class

Fig. 2. Age distribution per class (Hypothyroid vs Hyperthyroid)

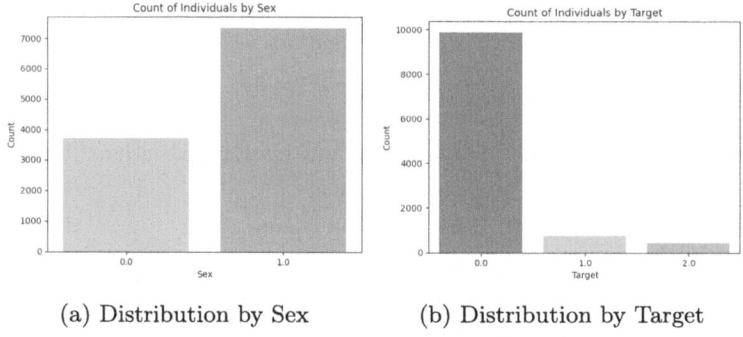

(a) Distribution by Sex (b) Distribution by Target

Fig. 3. Distribution by Sex and Target

The distribution of the class samples is depicted in Fig. 3b. It can be seen that the dataset is highly unbalanced - 9867 samples without disease (class 0), 753 hypothyroid patients (class 1) and 446 hyperthyroid patients (class 2).

In patients with hypothyroidism, TSH levels are usually high, T3 and T4 levels are low (particularly in severe cases), T4U may be low, normal or slightly high depending on the underlying cause and the stage of the disease, and FTI is generally low. The subplots in Fig. 4a corroborate with these medical evidences.

Similarly, the box-plots in Fig. 4b confirm that in patients with hyperthyroidism, TSH levels are usually below the normal reference range, T3 and T4 levels are generally increased, T4U may be normal or slightly elevated. Due to an increase in total thyroxine levels and the decrease in TSH levels, the FTI is elevated.

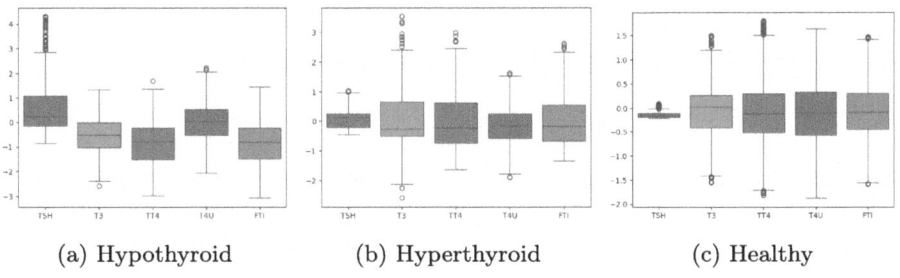

Fig. 4. Major features (TSH, T4, T3, T4U, FTI) box plots per class

4 Methodology

4.1 Feature Selection

During the preprocessing step samples with missing values and outliers were eliminated. Data was further standardized in order to bring the features to a common scale.

The next step was to assess the importance of the features and their correlation with the class. Following the ideas presented in [2,11], we have applied Machine Learning Feature Selection (MLFS), based on entropy criteria and decision tree classifier, to rank the features according to their contribution to the discrimination capacity of the models.

In Fig. 5a is presented the output of the feature importance evaluation using MLFS, and in Fig. 5b is presented the correlation between the features and the target class with feature importance value higher than the threshold of 0.015, as suggested in [2]. Following these results we considered two scenarios: thyroid disorder diagnosis based on the complete set of features and based only on the selected features (age, TSH, TT4, T4U, T3, and FTI).

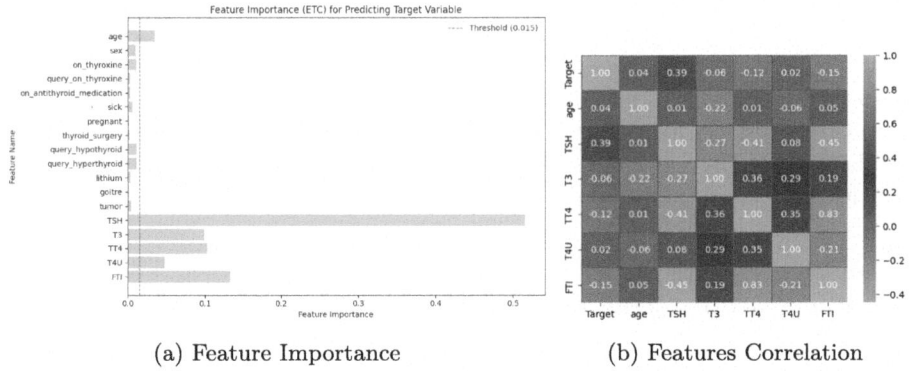

Fig. 5. Feature Importance and Correlation

4.2 Class Imbalance

The class imbalance issue (see Fig. 3b) was handled with the Synthetic Minority Oversampling Technique (SMOTE), that creates new samples for the underrepresented classes in the dataset, [12]. SMOTE selects a random data point on the feature space of the minority class and then identifies the k-nearest neighbors of that class, connecting these data points with virtual lines. Afterwards, new samples are randomly generated on the these virtual lines, [13].

4.3 Classification Models

The search for the best diagnosis configuration is based on previous state of the art related work.

Predicting thyroid disease conditions using DL models like DNN, CNN, LSTM, and a hybrid CNN-LSTM can reveal highly effective when applied to medical data. Each one of these models offer unique advantages during processing and analysis of complex data found thyroid medical records.

DNNs are very versatile and can be particularly useful for feature processing, and to capture complex and non-linear relations between features.

Despite of being typically used and associated with image processing, [14], CNNs can easily be adapted to disease prediction. These models are recognized for their ability to recognize patterns, and almost automatically extract the relevant features from the input data, potentially uncovering hidden patterns, [15].

LSTM are particularly useful for analyzing sequential or time-series data. They can capture long-term dependencies in hiden patterns within the data, can handle medical records of different lengths, and accommodate patients with different test frequencies.

The combination of CNN and LSTM models in a hybrid version offer several advantages. The complementary strengths of both models converge into an efficient feature learning, and improved accuracy.

The following models were trained and compared:

Model 1: Fully connected DNN with 3 hidden layers and 256, 28 and 63 Relu neurons respectively and softmax output layer was proposed in [16] for a binary classification of thyroid cancer. We use the same architecture with the adjustment for a three class problem.

Model 2: CNN with the following architecture: an embedding layer, 1D convolutional layer (Conv1D) with 128 filters and Relu activation function, a max-pooling layer, a flatten layer to convert the output to one dimensional vector and finally a softmax layer.

Model 3: LSTM with the following architecture: an embedding layer, a dropout layer (rate of 0.5) a LSTM layer (128 units) and a softmax layer.

Model 4: Hybrid architecture was also trained composed of CNN and LSTM with the same structures as in Model 2 and Model 3 respectively.

Model 2, Model 3 and Model 4 were inspired by the recent work of Chaganti et al., [2] and compared with their results in the next section. The combination of CNN and LSTM architectures for Model 4 was done to leverage the capabilities of CNN to extract and learn hidden patterns in the data, and the ability of LSTM to capture dependencies in the hidden patterns, obtaining a superior performance, compared to other common neural networks. Data was divided into training (80%) and test (20%) subsets. All models were trained with Adam optimizer and categorical cross entropy loss function [17]. Their performance was assessed based on the accuracy, recall, precision, F1-score (trade-off between precision and recall) metrics computed with the test data.

5 Results

In order to find the best predictive configuration, the four models, described in the previous section, were trained in the following scenarios: with all original (18) features or with the MLFS selected (6) features; with SMOTE class balancing or without class balancing. In total, 16 models were trained and tested.

All models were performed in a 13th Generation Intel Core i7-1355U 1,70 GHz with 16 Gb of memory RAM machine with a Windows 11 operation system. To implement all the proposed architectures, Keras, and Sci-kit learn frameworks were used with Python programming language.

We have chosen not to tune the architectures of the models in terms of number of layers, number of neurons, type of activation functions, and other topology hyper-parameters. Instead we have used previously developed models for similar thyroid pathology problem and concentrated on data centered aspects in order to enhance their performance.

Table 2. Performance metrics (with test data) of the best models

Models	Class Balancing	Features	Accuracy	Recall	F1-Score	Precision
DNN	SMOTE	All features	0.92	0.76	0.71	0.67
CNN	SMOTE	All features	0.99	0.93	0.94	0.95
LSTM	SMOTE	All features	0.99	0.94	0.94	0.94
CNN + LSTM	No class balancing	*age, TSH T3, TT4 T4U, FTI*	0.99	0.94	0.95	0.96

Table 3. Performance results in [2]

Models	Features	Accuracy	Recall	F1-Score	Precision
LSTM	All features	0.84	0.83	0.80	0.84
CNN	All features	0.93	0.94	0.92	0.94
CNN-LSTM	MLFS	0.92	0.91	0.91	0.91

Table 2 summarises the performance metrics of the best models according to the chosen evaluation metrics. Model 4 (CNN + LSTM), with the reduced set

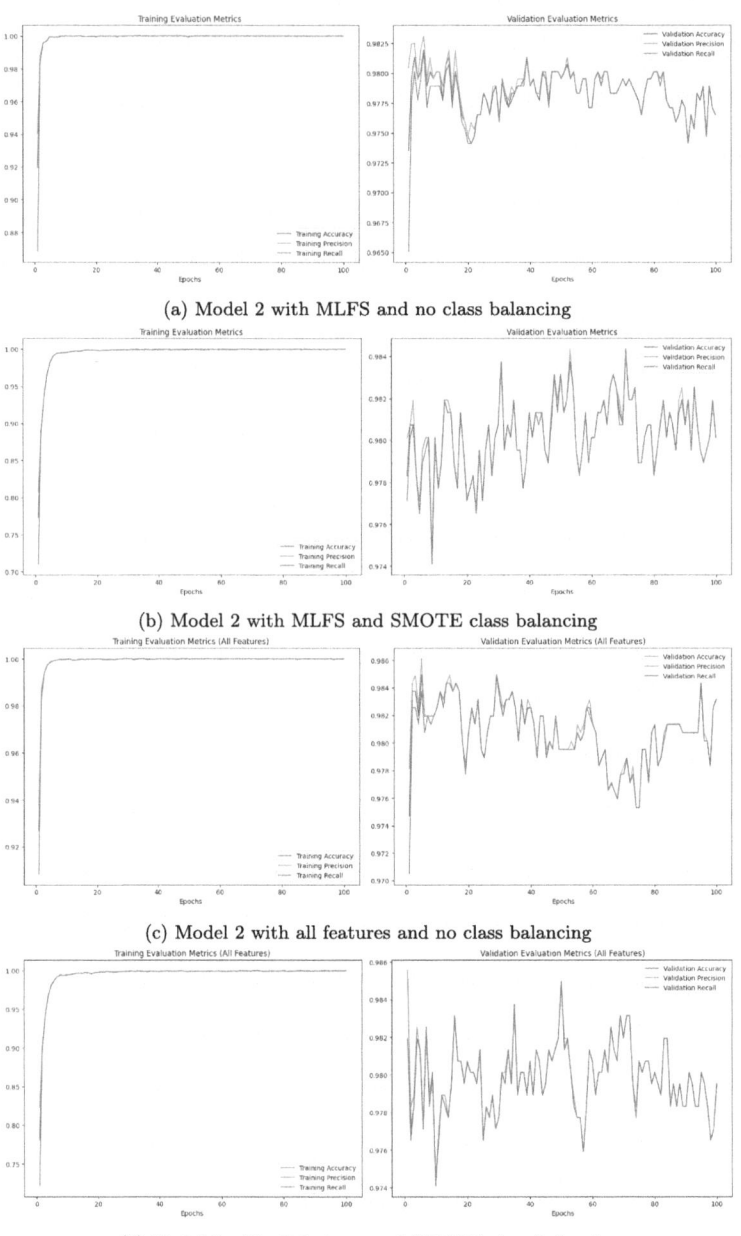

Fig. 6. Model 2 (CNN) training and validation scenarios

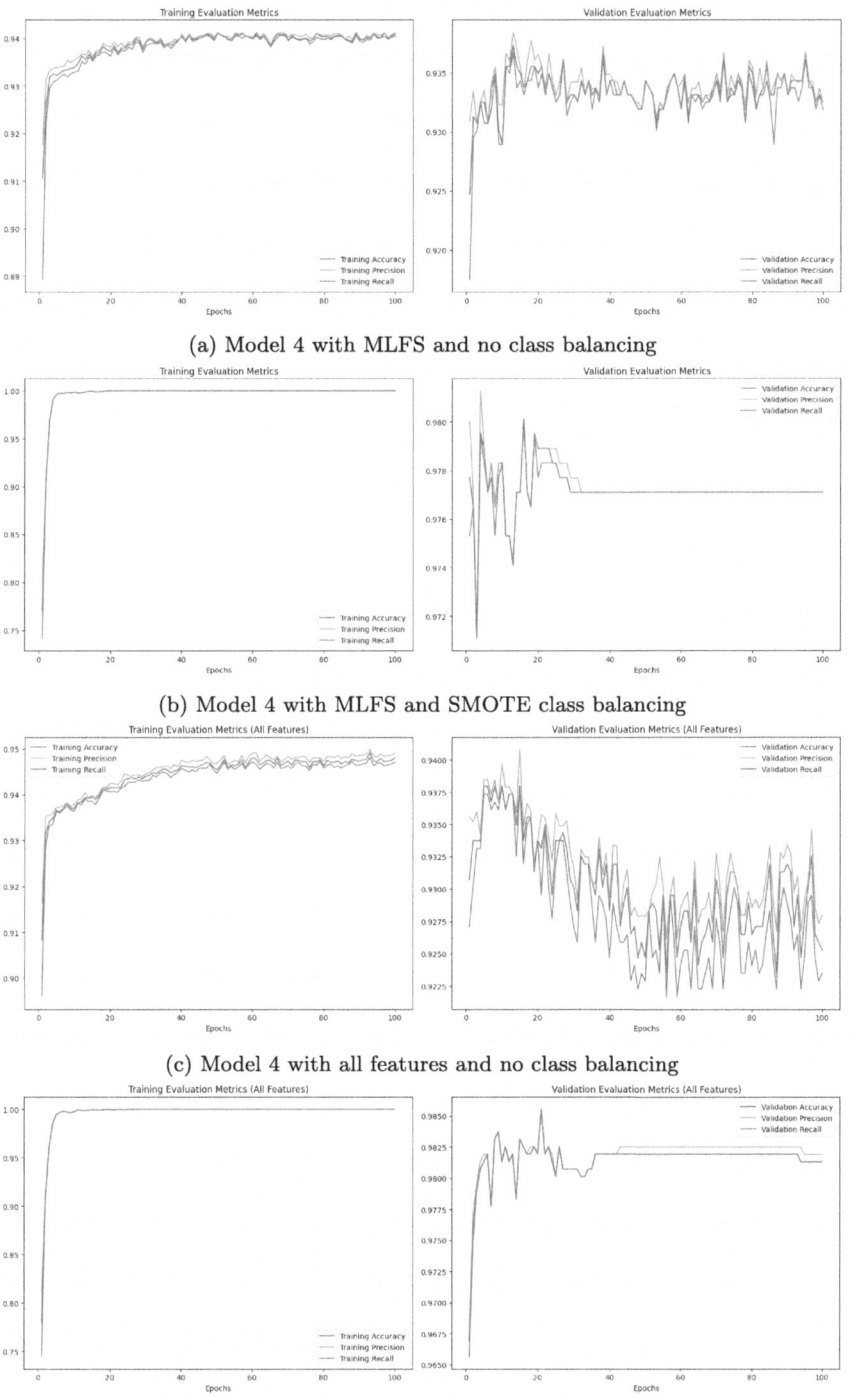

(a) Model 4 with MLFS and no class balancing

(b) Model 4 with MLFS and SMOTE class balancing

(c) Model 4 with all features and no class balancing

(d) Model 4 with all features and SMOTE class balancing

Fig. 7. Model 4 (CNN+LSTM) training and validation scenarios

of features (*age*, *TSH*, *T3*, *TT4*, *T4U* and *FTI*), slightly outperforms Model 2 (CNN) and Model 3 (LSTM) in the classification of hyperthyroidism, hypothyroidism disorders and healthy thyroid. This result is attributed to the feature selection step that eliminates features with marginal importance for the thyroid pathology. Model 1 (DNN) exhibit significantly worse performance, most probably because its architecture has been previously fitted for a binary classification, [16]. The performance metrics (Accuracy, Precision, Recall) obtained with training and validation data for all configurations of Model 2 and Model 4 are illustrated in Fig. 6 and Fig. 7, respectively. Persistent overfitting issues are observed in most of the scenarios, with exception of Model 4 (CNN + LSTM) with MLFS features and no class balancing (Fig. 7a). The obtained results demonstrate superior performance, compared to the state of the art work in [2], as shown in Table 3. SMOTE-based class balancing brought also potentially expected enhancement of the performance metrics.

6 Conclusions

Data-based approach for diagnosis of thyroid disorders is still at its early stage. Most of the published works deal with binary classification of the disorders, i.e. presence or not of some pathology (cancer, hyperthyroidism, hypothyroidism, etc.). The present work extends the problem to a multi-class diagnosis of typical thyroid disorders, namely hypothyroid, hyperthyroid and no pathology thyroid. The proposed DL models (DNN, CNN, LSTM, CNN+LSTM) were inspired by recently published results in [2] and [16]. Proper feature selection and the class balancing with SMOTE have favourable impact on the generalization properties of the models. All performance metrics (accuracy, recall, precision, F1-score) agree on high discrimination of the three classes.

Our findings suggest that DL models have the potential to diagnose different thyroid disorders and support healthcare professionals in clinical decision-making and ultimately contribute to improving patient care and thyroid disease management.

By leveraging machine learning, we can potentially revolutionize healthcare in several key areas. ML algorithms are able to analyze a considerable amount of patient data, potentially leading to more accurate diagnoses compared to traditional methods. This can reduce the risk of misdiagnosis and also improve patient outcomes. ML-empowered systems can also analyze data and suggest diagnoses much faster than traditional methods. This can lead the health system to a quicker treatment initiation and potentially better health outcomes for patients. Automating certain diagnostic tasks using ML can free up valuable time for healthcare professionals, enabling them to concentrate on more complex tasks and cases and deliver more personalized care to their patients. Finally, by analyzing large datasets, ML can easily identify patterns and trends in thyroid disorders, enabling healthcare providers to develop targeted prevention and screening programs.

Further steps of this study, are to add an explainable mechanism, that would be valuable for the medical professionals. Incorporating interpretability techniques help to understand how the models arrive to their decisions, leading to a more trustworthy and reliable diagnostic system.

Acknowledgments. This work was supported by the European Union - NextGenerationEU, through the National Recovery and Resilience Plan of the Republic of Bulgaria, project N0 BG-RRP-2.004-0005. The work is further funded by FCT/MCTES through national funds and when applicable co-funded EU funds under the project UIDB/50008/2020-UIDP/50008/2020.

References

1. American Thyroid Association. Thyroid function tests. Brochure (2019). Accessed 25 Mar 2024
2. Chaganti, R., Rustam, F., Díez, I., Mazón, J., Rodríguez, C., Ashraf, I.: Thyroid disease prediction using selective features and machine learning techniques. Cancers **14**, 3914 (2022)
3. Salman, K., Sonuç, E.: Thyroid disease classification using machine learning algorithms. J. Phys. Conf. Ser. **012140**(07), 2021 (1963)
4. Sapinho, I., Cid, M.: Doenças da tiroide: o que é, sintomas e tratamento. CUF (2023)
5. Jha, R., Bhattacharjee, V., Mustafi, A.: Increasing the prediction accuracy for thyroid disease: a step towards better health for society. Wireless Pers. Commun. **122**, 1–18 (2022)
6. Razia, S., SwathiPrathyusha, P., Krishna, N., Sumana, N.: A comparative study of machine learning algorithms on thyroid disease prediction. Int. J. Eng. Technol. **7**, 315 (2018)
7. Shahid, A., Singh, M., Raj, R., Suman, R., Jawaid, D., Alam, M.: A study on label TSH, T3, T4U, TT4, FTI in hyperthyroidism and hypothyroidism using machine learning techniques, pp. 930–933 (2019)
8. Riajuliislam, M., Rahim, K., Mahmud, A.: Prediction of thyroid disease(hypothyroid) in early stage using feature selection and classification techniques. In: 2021 International Conference on Information and Communication Technology for Sustainable Development (ICICT4SD), pp. 60–64 (2021)
9. Naeem, A., et al.: Hypothyroidism disease diagnosis by using machine learning algorithms. Int. J. Intell. Syst. Appl. Eng. **11**(3), 368–373 (2023)
10. Quinlan, R.: Thyroid Disease. UCI Machine Learning Repository (1987). https://doi.org/10.24432/C5D010
11. Akash, K.T., Usman, F.M., Kumar, T.N., Ahmed, M.R., Gudodagi, R.: Predicting thyroid dysfunction using machine learning techniques. In: 2023 12th International Conference on Advanced Computing (ICoAC), pp. 1–8 (2023)
12. Blagus, R., Lusa, L.: SMOTE for high-dimensional class-imbalanced data. BMC Bioinform. **14**(1) (2013)
13. Brownlee, J.: Smote for imbalanced classification with python (2017)
14. Hernandez, D., Pereira, R., Georgevia, P.: Covid-19 detection through x-ray chest images. In: 2020 International Conference Automatics and Informatics (ICAI), pp. 1–5. IEEE (2020)

15. Georgieva, O., Milanov, S., Georgieva, P., Santos, I.M., Pereira, A.T., Silva, C.F.: Learning to decode human emotions from event-related potentials. Neural Comput. Appl. **26**, 573–580 (2015)
16. Pandey, A., Barve, A.: Using deep neural network to detect thyroid cancer. In: 2022 11th International Conference on System Modeling & Advancement in Research Trends (SMART), pp. 1347–1351. IEEE (2022)
17. Ho, Y., Wookey, S.: The real-world-weight cross-entropy loss function: modeling the costs of mislabeling. IEEE Access **8**, 4806–4813 (2020)

Advanced CNN-SVM Machine Learning Techniques for Facial Skin Ultrasound Image Analysis

Aayad Nabeel[1], Mostafa Ragheb[2], Galina Momcheva[3](✉), Issa Kamar[4], and Mohamad Hamady[4]

[1] Al-Mansour University College, Baghdad, Iraq
[2] Faculty of Sciences, Lebanese University, Tripoli, Lebanon
[3] Institute of Mathematics and Informatics, Bulgarian Academy of Sciences, Sofia, Bulgaria
gmomcheva@math.bas.bg
[4] Arts Sciences and Technology University in Lebanon, Beirut, Lebanon

Abstract. This study investigates the machine learning techniques for unsupervised image classification and quality assessment in the domain of ultrasound imaging. Leveraging Convolutional Neural Networks (CNNs) for feature extraction and subsequent integration into a Support Vector Machine (SVM) model, we explored a novel approach aimed at accurate image classification. The dataset comprises high-frequency images in the form of image sequences depicting the facial skin of females. The study's primary emphasis was to categorize ultrasound images based on learned deep features, offering a distinctive framework for unsupervised image classification. The investigation employed CNNs to extract deep features from images, enhancing the SVM model's performance in accurately categorizing images. The incorporation of gamma correction as a preprocessing step further augmented the accuracy and sensitivity of the models. The SVM model exhibited exceptional performance, achieving accuracy rates exceeding 95.43% in the training phase and approximately 94.72% during testing, that is a significant milestone in the precise classification of ultrasound images.

Keywords: medical ultrasound · deep learning · facial skin images · CNN-SVM

1 Introduction

Analyzing face in every aspect from computer vision point of view is a current and prospective area of research for the purpose of wellbeing, cosmetic industry, biotechnology industry and biometric security. In some recent studies the approach of CNN-SVM is used for automatically detecting and diagnosing dermatological problems and their diagnostics, which has the potential to change clinical procedures and enhance patient outcomes in the area of skincare therapy [1].

In cosmetic research, skin can be classified into four main categories as, normal, dry, oily and combination. The current methods to identify the cosmetic skin type are time consuming and error prone [2].

During lengthy and high-intensity clinical ultrasound examinations [3], it can be challenging for clinicians to consistently guarantee the quality of ultrasound images. Noise, shadows, low contrast, and other sensory and motion artifacts make ultrasound pictures challenging to comprehend. The fuzzy borders and presence of different anatomical elements make assuring quality during acquisition much more challenging. As a result of the need for low human effort and uniform quality among beginner sonographers during picture capture, and for ultrasound image quality assessment (US-IQA), the research community has embraced automation [4].

The existing literature has extensively explored different machine learn-ing techniques, including supervised, unsupervised, and deep learning approaches. In this particular study, we will delve into the effectiveness of transfer learning and CNN in evaluating ultrasound images. As part of the investigation, we apply gamma correction to the images as a pre-processing step and eliminate periodic noise in the images. Furthermore, we compare the performance of implementing CNN both with and without the pre-processing stage. By carrying out this study, we hope to better understand how these methods aid in the evaluation of ultrasound images and ascertain how pre-processing affects CNN efficiency.

In a systematic review article (Vatiwutipong) the existing research applications of AI in the field of cosmetic dermatology is analyzed. It gives trends in AI research and application in aesthetic medicine and guidance for practitioners seeking to implement AI technologies to address real-world challenges in cosmetic services [5]. The results of our article have contribution to the same area of research and innovative practices.

2 Related Work

Recently a research about applications of Convolutional Neural Network for skin type classification have been studied by Kothari. The results of our CNN classification model show an accuracy of about 85% with a slight bias towards oily images. The results show that Deep Learning has great potential in the field of skin type classification from facial images, and with a dataset of greater size, could give more optimal and even lesser error prone results [2]. Other researchers Mishra investigate several methods for classifying skin illnesses [6], where Hadi did comparison between CNN and SVM in the area of skin cancer images recognition [7]. An interesting study is published by Kim for facial expression recognition using hybrid CNN and SVM that is useful for effective human–computer interaction, robot interfaces, and emotion-aware smart agent systems [17].

Numerous unsupervised learning algorithms for medical imaging analysis have been looked into in order to get around the issues with labor-intensive annotation in guided learning [8]. In addition to modern techniques like Autoencoder (AE), generative networks [9], and Deep Clustering [10], traditional clustering approaches were used. One of the most effective methods for compressing low-dimensional latent space and extracting unsupervised features while minimizing re-creating error between encoded input and decoded output is the AE technique. They are primarily used in medical images to detect anomalies and denoising [11]. Furthermore, for the analysis of medical images, generative networks like Variational AE (VAE) have demonstrated considerable promise

[12]. They impose a predetermined distribution in the latent space during training using variational loss. Nesovic and colleagues [13] combined AE and a Random Forest Classifier (RFC) to address the US-IQA. RFC, however, requires noise-specific handmade characteristics, which are difficult to generate for noisy ultrasound scans. Clustering is another effective unsupervised strategy for detecting patterns in unlabeled datasets. Uniform Manifold Approximation and Projection (UMAP) [14] and Principal Component Analysis (PCA), Spatial Clustering using Density of Utilization with Noise (DBSCAN) [15], the spectral clustering method [16], k-means are examples of traditional clustering algorithms. They, however, are unable of dealing with high-dimensional data. Deep clustering has lately piqued the attention of unsupervised image analysis researchers [18]. By fusing AEs and VAEs with common clustering algorithms, deep cluster learning learns the clusterable representation in latent space. This method has been used to categorize handwritten numbers, face images [19], MRI imaging [20], and CT images. In images Yu et al. [21] demonstrated deep clustering for the detection of thyroid cancer. Notably, de-spite the fact that it is challenging to examine due to noise, artifacts, different perspectives, low interclass and high intraclass variability, deep clustering for US-IQA has not yet been discovered. Breast ultrasound images do not plainly show breast anatomical structures like the pericardium, subcutaneous fat layer, or other structures, and because each person is unique, their breast morphologies differ significantly. It is therefore difficult to evaluate the quality of breast ultra-sound images by removing and identifying specific anatomical functions. Song and others learned the clinicians' scientific assessment criteria by using the DL approach to evaluate the quality of medical breast ultrasound images [22]. Skilled experts evaluated the 1205 breast ultrasound images that were gathered from 533 people for the study. The ResNet18, which has few layers, is used to extract features from ultrasound images, and the effective characteristic coding model Bilinear CNN, which uses fine-grained classification to improve quality evaluation performance. The Pearson Linear Correlation Coefficient value was 0.842 for the consistency between the results of our method's ultrasound picture quality evaluation and the annotations of the doctors [23].

3 Methods and Materials

3.1 Proposed Approach

The section presents a methodology proposed for evaluating the quality of ultrasound images of facial skin, employing gamma correction, a CNN network, and an SVM classifier. This approach, depicted in Fig. 1, begins with loading the dataset of scored images. Subsequently, the samples are divided, with 80% allocated for training and validation, and 20% for testing. The US images are then processed and classified through several steps: loading the images, applying gamma correction, building and training the CNN network, extracting new features with CNN, reapplying gamma correction to the new deep features array, and classifying the features with SVM. Finally, the classification performance is evaluated using the testing samples, with a particular focus on examining the differences between using CNN as both a classifier and a feature extraction tool.

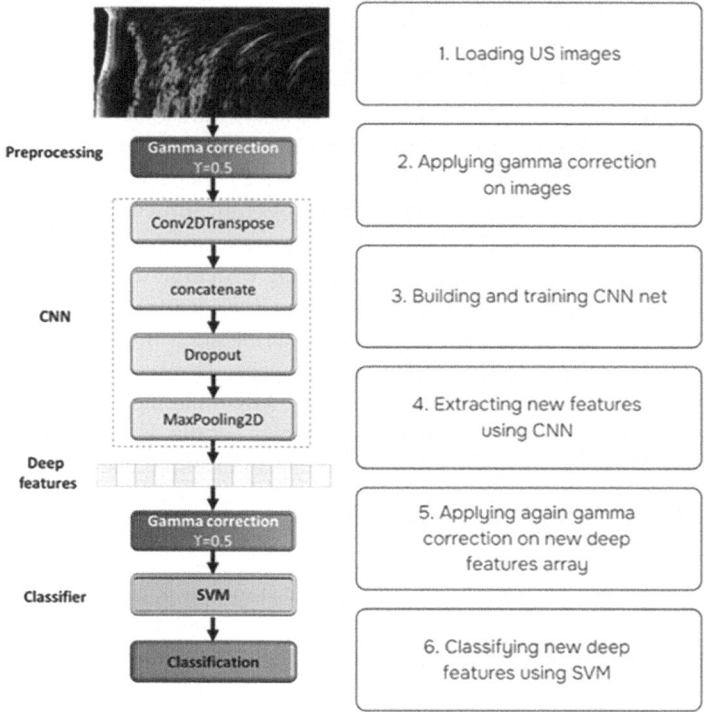

Fig. 1. The proposed and implemented methodology for US IQA

3.2 Dataset

The dataset comprises high-frequency images in the form of image sequences depicting the facial skin of females. This dataset was gathered over four sessions, identified by data IDs in the format [day month year]. The study involved 44 healthy Caucasian female subjects aged between 56 and 67 (mean = 60.64, standard deviation = 2.61), all in the postmenopausal phase. As part of an **anti-aging skin therapy**, subjects underwent treatment involving a trichloroacetic acid (TCA) chemical peel. The initial image data were obtained before the first application of the acid, and participants were divided into treated (23 subjects) and placebo (21 subjects) groups.

The data were captured from three distinct locations on the facial area, with the movement direction of the ultrasound probe depicted in Fig. 1 using superimposed arrows on a facial model. The image acquisition commenced where the arrow started and concluded at the arrow's end. During each patient visit, three high-frequency ultrasound (HFUS) series were recorded. Each series comprised approximately 40 HFUS images for each location. The original image resolution stood at 1386 × 3466 pixels, with a pixel size of 0.0093 × 0.0023 mm/pixel (axial × lateral).

The HFUS image data were collected using the DUB SkinScanner75, employing a 24 MHz transducer operating at a B-mode frequency, 8 mm depth, and an acoustic intensity level of 40 dB. Each series encompassed both technically feasible image data for further diagnosis (utilizing CAD or medical software) and data considered unsuitable for analysis. For instance, the latter category included ultrasound frames captured with inadequate probe adherence to the patient's skin or when the angle between the probe and the skin was <70 degrees. Example HFUS images classified as 'ok' (suitable for analysis) or 'no ok' (not suitable for analysis) are displayed in Fig. 1.

The HFUS examinations were conducted by two sonographers, one with novice experience in HFUS (with no prior image acquisition experience but working with conventional ultrasound) and an experienced sonographer trained at the Euroson School Sono-Derm, specializing in HFUS image analysis for three years. A total of 17,425 HFUS images were obtained.

Following data collection, the entire dataset underwent annotation by two HFUS data analysis experts. One expert provided annotations twice with a one-week interval. As a result, the subsequent descriptions are categorized into three annotations—Expert 1, Expert 2, and Expert 3. It is noteworthy that labels Expert 1 and Expert 2 pertain to the same individual (annotations by the first expert with a one-week interval) as shown in Fig. 2.

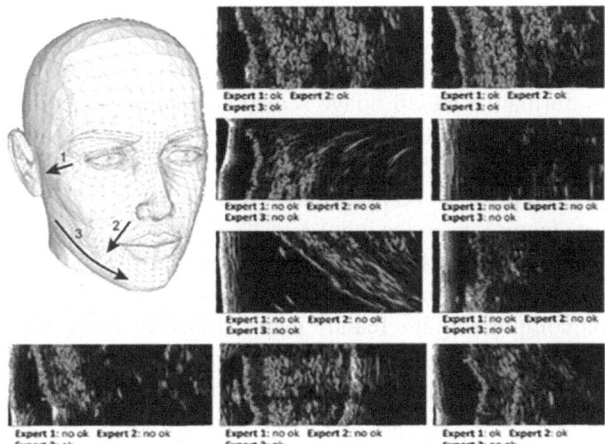

Fig. 2. Expert-annotated facial model with overlay image capture regions and example HFUS scans [23].

3.3 Gamma Corrections

Gamma correction is also used to rectify any inconsistencies in color intensity or brightness, particularly in medical imaging like US scans, especially in specific areas of interest (as shown in Fig. 3). In this process, the input value is raised to the power of the inverse of gamma to calculate the gamma correction.

The formula used for gamma correction is typically expressed as follows:

$$I\prime = 255 \cdot \left(\frac{I}{255}\right)^{\gamma} \tag{1}$$

Where the gamma value determines the degree of correction applied to the in-put values to ensure that the colors and intensities are appropriately represented in the final image displayed on the screen. In this study, ($\Upsilon = 0.5$) is used in implementing gamma correction.

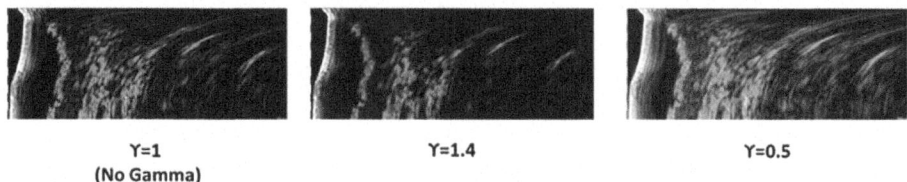

Y=1 (No Gamma) Y=1.4 Y=0.5

Fig. 3. US images with different values of gamma correction.

3.4 CNN Model

The first goal of this step is the CNN model, providing the most reliable classification results. Based on the previous experiences [18–23] and the recent papers in medical IQA [18–23], or informative HFUS image selection, we consider the following CNN parameters: The proposed Convolutional Neural Network (CNN) model incorporates an extensive architecture, specifically designed with 20 convolutional layers. The initial layer of the network comprises 64 convolution kernels each of size 5x5, serving to extract and capture intricate features from the input data. The stage size is set to 1 x 1, ensuring a detailed evaluation of the data at the initial stage. Additionally, a 2 x 2 maximum pooling layer is integrated into the architecture. This layer is crucial for discarding low-level local characteristics and compressing the feature maps, streamlining the subsequent layers' evaluation. In this network, two activation layers, employing the 'relu' (Rectified Linear Unit) activation function, play a pivotal role in introducing non-linearity and enhancing the model's capability to learn complex patterns and relationships within the data.

Before training, the RGB ultrasound frames underwent resizing to dimensions of 224 × 224 × 3 pixels. Through a series of experiments, it was determined that the stochastic gradient descent optimizer, configd with a momentum of 0.9, utilization of categorical cross-entropy as the loss function, a batch size set at 64, and an initial learning rate of 0.001, proved to be the most efficient configuration.

3.5 Using CNN as Feature Extraction Tool

The approach is designed to harness learned image features acquired from a trained CNN and leverage these features to train an SVM classifier. The proposed method (refer to Fig. 1) encompasses a series of steps: (1) the establishment and training of a CNN utilizing the dataset; (2) identification of the fully connected layer within the CNN; (3) extraction of feature representations from this layer, containing the acquired features; (4) construction of an SVM classifier model based on the training dataset; (5) an assessment of the CNN and SVM performance employing the extracted features from the testing set.

3.6 Using CNN as Feature Extraction Tool

The SVM classifier with a radial kernel, often referred to as the Radial Basis Function (RBF) kernel, is a powerful algorithm for classification tasks using deep features derived from networks like CNNs. Once trained, the SVM can classify new or unseen data based on the patterns and characteristics it has learned from the deep features. The SVM classifier with a radial kernel operates on deep features derived from CNNs, enabling effective classification by exploiting the complex patterns and high-level representations learned by the CNN. This approach is particularly beneficial for handling non-linear and complex data distributions where traditional linear classifiers might not be sufficient.

4 Results

4.1 Training CNN Model

The comprehensive performance analysis of the training procedures has been meticulously outlined in Table 1 within the context of the academic article. Initially, without the implementation of GC in the preprocessing phase, the accuracy achieved during the training of the CNN was determined to be 81.34%. This accuracy metric encapsulates the network's overall ability to make correct predictions. Additionally, two essential evaluation metrics, sensitivity, and F1-score, were also quantified. The sensitivity, also known as the true positive rate, stood at 79.99%, indicating the network's proficiency in accurately identifying positive instances within the dataset. Meanwhile, the f1-score, a harmonic mean of precision and recall, was determined to be 78.98%, serving as a robust evaluation of the network's overall accuracy in classifying data. A noteworthy improvement was observed upon the inclusion of GC as a preprocessing step for all US images. This augmentation distinctly enhanced the performance outcomes during train-ing. With GC employed in the preprocessing stage, the CNN achieved a substantial rise in accuracy, escalating to an impressive 93.12%. Furthermore, the sensitivity metric experienced a significant increase, elevating to a notable 92.22%.

Table 1. The performance of testing the CNN model as a classifier.

Heading level	Training without GC	Training with GC
Accuracy (%)	81.34%	93.12%
Sensitivity (%)	79.99%	92.22%
F1-score (%)	78.98%	91.99%

4.2 The Outcomes of Proposed Method

Some details of the training and testing procedures are comprehensively showcased in Table 2, offering a detailed evaluation of the model's performance. During the training phase, the SVM model showcased very good results, achieving an accuracy surpassing 95.43%. This robust performance continued through the testing phase, where the SVM model achieved an impressive accuracy rate of 94.72%. This accuracy metric encapsulates the model's capability to accurately classify and discern patterns within the dataset. Moreover, the F1-score, exceeding 92%, serves as an essential evaluation metric indicating the model's ability to minimize false positive outcomes. This high F1-score underlines the SVM classifier's proficiency in achieving a low number of erroneous positive classifications, reinforcing its reliability in accurately discerning and categorizing features within the dataset.

Table 2. The performance of training and testing SVM classifier.

Heading level	Training		Testing	
	Ok	Not Ok	Ok	Not Ok
Accuracy (%)	95.43		94.72	
Sensitivity (%)	94.65	96.94	93.22	92.43
F1-score (%)	94.23	96.12	95.45	93.18

5 Discussion

In comparison to the array of DL algorithms and methodologies described in various studies, the performance detailed in the previously mentioned SVM model signifies a noteworthy advancement in supervised strategies for image classification. While traditional clustering algorithms such as UMAP, PCA, DBSCAN, spectral clustering, k-means, and t-SNE offer valuable methods for pattern detection in unlabeled datasets, their limitation in dealing with high-dimensional data is evident. This limitation contrasts with the SVM model's robustness, achieving an accuracy exceeding 95.43% in training and 94.72% in testing. The SVM model also excels in sensitivity, surpassing 93% for classifying 'good' US images and approximately 95% for 'normal' images, whereas the mentioned traditional clustering algorithms face challenges with high-dimensional data. Moreover, the described SVM model's F1-score, exceeding 92%, showcases its efficiency in minimizing false positive outcomes. The integration of deep features extracted from the CNN, combined with the preprocessing step of GC, enhanced the classification accuracy substantially, marking a significant stride in accurate image classification.

In comparison to the innovative studies focusing primarily on the evaluation of ultrasonography image quality in terms of anatomical features and specific anatomical components, the SVM model's emphasis on image classification pre-sents a distinct objective. Zhang and colleagues proposed an automated image quality evaluation system based on multi-task learning, aiming to assess the completeness, clarity, and visibility of anatomical features within ultrasonography images. Similarly, Lin and team suggested evaluating the quality of fetal head ultrasound images using a Multitask Fast

(MFR)-CNN, which aimed to locate and identify vital anatomical components. However, the primary objective of these methodologies was to assess image quality based on anatomical visibility and completeness rather than unsupervised categorization. Moreover, evaluating breast ultrasound images poses significant challenges due to the unique morphologies of breast structures among individuals, making it difficult to assess quality by identifying specific anatomical functions. Song and collaborators ad-dressed this by using a DL approach to assess the quality of medical breast ultra-sound images. Their approach utilized ResNet18 and a Bilinear CNN model, aim-ing to extract features and enhance quality evaluation performance through fine-grained classification. The Pearson Linear Correlation Coefficient value obtained, indicating consistency between the method's quality evaluation and ex-pert annotations, was 0.842. In contrast, the SVM model described in the original context focused on the classification of images using deep features obtained from CNNs, incorporating a distinctive approach that aimed to categorize images rather than evaluate image quality based on specific anatomical components.

6 Conclusion

The study's outcomes and subsequent discussions shed light on the considerable advancements in image classification achieved by leveraging CNN and SVM models. The use of deep features extracted from CNNs, coupled with innovative preprocessing steps such as GC, significantly enhanced the accuracy and sensitivity of the models. The SVM model, in particular, demonstrated exceptional performance, achieving accuracy rates exceeding 95.43% in training and around 94.72% in testing. This marked success signifies the robustness of utilizing deep learning techniques for image classification, providing an efficient and accurate framework for unsupervised categorization. The comparison with various studies emphasizing image quality evaluation based on anatomical completeness and visibility underscores the unique contribution of the present research. While those methodologies excel in assessing image quality through DL approaches and specific anatomical feature identification, the emphasis of the current study on image classification and pattern recognition presents a distinctive and effective approach.

The future of US image quality assessment could significantly benefit from the integration of these models. Further research could focus on the development of specific models geared toward assessing ultrasound images quality based on anatomical visibility and completeness. The proposed SVM model, augmented with deep learning techniques and GC preprocessing, could provide a foundation for an innovative framework aimed at evaluating the quality of ultrasound images. Additionally, exploring hybrid models that merge the prowess of deep learning for feature extraction with traditional image quality assessment methodologies could offer a comprehensive solution. These models could integrate the precise classification abilities of CNN-based feature extraction with the context-driven assessment provided by traditional quality evaluation methods.

Disclosure of Interests. The authors have no competing interests to declare that are relevant to the content of this article.

References

1. Mir, T.A., Banerjee, D., Upadhyay, D., Rawat, R.S.: Comprehensive facial acne classification: CNN-SVM synergy. In: 2nd World Conference on Communication and Computing (WCONF), RAIPUR, India, pp. 1–5 (2024)
2. Kothari, A., Shah, D., Soni, T., Dhage, S.: Cosmetic skin type classification using CNN with product recommendation. In: 12th International Conference on Computing Communication and Networking Technologies (ICCCNT), Kharagpur, India, pp. 1–6 (2021)
3. Carovac, A., Smajlovic, F., Junuzovic, D.: Application of ultrasound in medicine. Acta Informatica Medica **19**(3), 168 (2011)
4. Antico, M., et al.: Deep learning for US image quality assessment based on femoral cartilage boundary detection in autonomous knee arthroscopy. IEEE Trans. Ultra. Ferroelec. Frequency Control **67**(12), 2543–2552 (2020)
5. Vatiwutipong, P., Vachmanus, S., Noraset, T., Tuarob, S.: Artificial intelligence in cosmetic dermatology: a systematic literature review. IEEE Access **11**, 71407–71425 (2023)
6. Mishra, S., et al.: A comprehensive review on skin disease classification using convolutional neural network and support vector machine. In: Shaw, R.N., Paprzycki, M., Ghosh, A. (eds.) Advanced Communication and Intelligent Systems. ICACIS 2022. Communications in Computer and Information Science, vol. 1749. Springer, Cham (2023). https://doi.org/10.1007/978-3-031-25088-0_64
7. Zaid, G.H., et al.: Comparison between Convolutional Neural Network CNN and SVM in skin cancer images recognition. J. Tech. **3**(4), 15–22 (2021)
8. Raza, K., Singh, N.K.: A tour of unsupervised deep learning for medical image analysis. Curr. Med. Imaging **17**(9), 1059–1077 (2021)
9. Wang, G., Jiang, C., Shen, Z., Miao, Y., Wang, H.: SFGAN: unsupervised generative adversarial learning of 3D scene flow from the 3D scene self. Adv. Intell. Syst. **4**(4), 2100197 (2022)
10. Zhou, S., et al.: A comprehensive survey on deep clustering: Taxonomy, challenges, and future directions. arXiv:2206.07579 (2022)
11. Xu, J.: A review of self-supervised learning methods in the field of medical image analysis. Int. J. Image Graph. Sig. Process. (IJIGSP) **13**(4), 33–46 (2021)
12. Wei, R., Mahmood, A.: Recent advances in variational autoencoders with representation learning for biomedical informatics: a survey. IEEE Access **9**, 4939–4956 (2020)
13. Nesovic, K., Koh, R.G., Sereshki, A.A., Zadeh, F.S., Popovic, M.R., Kumbhare, D.: Ultrasound Image Quality Evaluation using a Structural Similarity Based Autoencoder. In: 2021 43rd Annual International Conference of the IEEE Engineering in Medicine and Biology Society (EMBC), pp. 4002–4005 (2021)
14. Ghojogh, B., Crowley, M., Karray, F., Ghodsi, A.: Uniform Manifold Approximation and Projection (UMAP). In: Elements of Dimensionality Reduction and Manifold Learning, pp. 479–497 (2023)
15. Verma, M., Srivastava, M., Chack, N., Diswar, A.K., Gupta, N.: A comparative study of various clustering algorithms in data mining. Int. J. Eng. Res. Appl. (IJERA) **2**(3), 1379–1384 (2012)
16. Xu, H., et al.: A spectral clustering method combining path with density. In: 2012 IEEE International Conference on Robotics and Biomimetics (ROBIO), pp. 695–698 (2012)
17. Kim, J.-C., Kim, M.-H., Suh, H.-E., Naseem, M.T., Lee, C.: Hybrid approach for facial expression recognition using convolutional neural networks and SVM. Appl. Sci. **12**(5493), 12115493 (2022)
18. Caron, M., Bojanowski, P., Joulin, A., Douze, M.: Deep clustering for unsupervised learning of visual features. In: Proceedings of the European conference on computer vision (ECCV), pp. 132–149 (2018)

19. Song, C., Liu, F., Huang, Y., Wang, L., Tan, T.: Auto-encoder based data clustering. In: Progress in Pattern Recognition, Image Analysis, Computer Vision, and Applications: 18th Iberoamerican Congress, CIARP 2013, Havana, Cuba, 20–23 November, Proceedings, Part I vol. 18, pp. 117–124 (2013)
20. Soleymani, F., Eslami, M., Elze, T., Bischl, B., Rezaei, M.: Deep variational clustering framework for self-labeling large-scale medical images. Med. Imaging Image Process. **12032**, 68–76) (2022)
21. Yu, R., et al.: Feature discretization-based deep clustering for thyroid ultrasound image feature extraction. Comput. Biol. Med. **146**, 105600 (2022)
22. Song, Y., et al.: Medical ultrasound image quality assessment for autonomous robotic screening. IEEE Robot. Autom. Lett. **7**(3), 6290–6296 (2022)
23. Czajkowska, J., Juszczyk, J., Piejko, L., Glenc-Ambroży, M.: High-frequency ultrasound dataset for deep learning-based image quality assessment. Sensors **22**(4), 1478 (2022)

Testing the NEAT Algorithm on a PSPACE-Complete Problem

Angel Marchev Jr.[1,2](✉), Dimitar Lyubchev[1], and Nikolay Penchev[2]

[1] University of National and World Economy, 19 December 8 Street, 1700 Sofia, Bulgaria
angel.marchev@unwe.bg
[2] Sofia University "St. Kliment Ohridski", 125 Tsarigradsko Shose Blvd., bl.3, 1113 Sofia, Bulgaria

Abstract. This paper investigates the efficacy of the Neuro-Evolution of Augmenting Topologies (NEAT) algorithm on PSPACE-complete problems, specifically utilizing the Sokoban puzzle. NEAT, which evolves both neural network topologies and weights, provides a promising approach for solving complex problems without predefined network architectures. We implemented NEAT using the neat-python library and tested it against several reinforcement learning (RL) algorithms, including Deep Q-Network (DQN) and Proximal Policy Optimization (PPO), within the OpenAI gym-sokoban environment. Our experiments involved extensive configuration variations to identify optimal settings for NEAT. Key findings indicate that NEAT solved the Sokoban problem, outperforming traditional RL variants. Our results highlight the importance of incremental structural growth and the protection of topological innovations. This study confirms NEAT's applicability to PSPACE-complete problems.

Keywords: Neuro-Evolution of Augmenting Topologies · PSPACE-Complete · Reinforcement learning

1 Introduction

Artificial Neural Networks (ANNs) are pivotal in modern computational approaches, enabling advancements across numerous domains. However, the challenge of defining optimal network architectures often necessitates prior domain knowledge, creating barriers for zero-knowledge architecture searches. This study addresses the need for architecture-independent methods by focusing on the NEAT (Neuro-Evolution of Augmenting Topologies) algorithm, as proposed by Stanley and Miikkulainen [1]. Despite its potential, the literature lacks comprehensive investigations into NEAT's efficacy on complex, PSPACE-complete problems. This paper aims to fill that gap by testing NEAT on the Sokoban problem, a known PSPACE-complete task, to evaluate its performance without predefined network structures.

This research employs an extensive testing methodology to explore NEAT's capabilities, leveraging a sophisticated dataset and rigorous experimental design. By comparing NEAT against various Reinforcement Learning (RL) algorithms, we demonstrate the broader applicability of our findings. The use of a challenging problem like Sokoban, combined with multiple configurations and robust checks, ensures the relevance and robustness of our results. This approach not only answers the specific research question but also provides insights into the generalizability of NEAT for complex problem-solving.

Our findings highlight the critical role of fitness customization, species diversity, and the careful management of network complexity in optimizing NEAT's performance. Specifically, NEAT Configuration 1 successfully solved the Sokoban problem within 1000 iterations, showcasing the effectiveness of the proposed adjustments. In contrast, traditional RL variants like DQN, PPO, PPO optimized, and PPO CNN did not achieve success, underscoring potential limitations in their configurations for this specific task.

Through sensitivity analyses and robustness checks, we validated the reliability of our results. Variations in initial nodes, fitness functions, population sizes, mutation rates, and training environments were tested to ensure comprehensive evaluation. NEAT Configuration 1's success demonstrates the viability of our proposed modifications, while the success of Q-learning in solving the problem suggests its potential applicability. Other RL variants' failures indicate areas for further investigation and improvement.

This study builds upon foundational work in neuroevolution, particularly the seminal paper "Efficient Evolution of Neural Network Topologies" by Kenneth O. Stanley and Risto Miikkulainen. By extending NEAT's application to a PSPACE-complete problem, we contribute to the broader literature on ANN architecture optimization and problem-solving without prior domain knowledge. Our work intersects with studies on evolutionary algorithms, reinforcement learning, and complexity theory, providing a comprehensive perspective on NEAT's capabilities.

We show that NEAT can effectively solve PSPACE-complete problems like Sokoban without predefined network structures. Our study underscores the importance of tailored fitness functions in improving neuroevolutionary outcomes. We highlight how maintaining species diversity and managing network complexity can lead to more robust solutions. By comparing NEAT with various RL algorithms, we provide a thorough evaluation of their respective strengths and limitations.

This paper is organized in several sections:

- **Introduction:** Establishes the research question and its significance.
- **Testing Methodology:** Details the experimental design and evaluation criteria.
- **Simulation Problem Definition:** Describes the OpenAI gymnasium environments and the gym-sokoban game.
- **Baseline Solution: Reinforcement Learning:** Discusses the implementation and performance of RL algorithms.

- **Neuro-Evolution of Augmenting Topologies:** Explains the NEAT algorithm.
- **Implementation of NEAT:** Explains the NEAT software implementation, setup and configurations.
- **Key Findings and conclusions:** Presents the findings from the experiments, summarizes the main insights, and implications and suggests potential areas for further exploration and development.

2 Testing Methodology

2.1 Using an API Game for Simulating the PSPACE-Complete Problem

To simulate the PSPACE-complete problem, we utilized the gym-sokoban environment from the OpenAI Gym third-party environments. Sokoban, a well-known PSPACE-complete puzzle game, provides a challenging testbed for evaluating the capabilities of neuroevolutionary algorithms like NEAT. The game's complexity and need for strategic planning make it an ideal candidate for this study.

2.2 Devise a Baseline Model

As a baseline model, we selected several popular Reinforcement Learning (RL) algorithms known for their applicability to similar types of problems. These included:

- **Deep Q-Network (DQN):** A widely-used algorithm that combines Q-learning with deep neural networks.
- **Proximal Policy Optimization (PPO):** A robust RL algorithm that balances exploration and exploitation.
- **Optimized PPO (PPO opt):** An enhanced version of PPO with improved hyperparameter tuning.
- **PPO with Convolutional Neural Networks (PPO CNN):** A variant of PPO that utilizes CNNs for better feature extraction.
- **Q-learning:** A traditional RL algorithm used as a comparative benchmark.

These baseline models were implemented and trained to solve the Sokoban problem, providing a standard against which to measure NEAT's performance.

2.3 Reproduce the Original Paper Using the Original Library

We reproduced the methodology outlined in Stanley and Miikkulainen's original NEAT paper using the neat-python library. This involved implementing the NEAT algorithm with its default settings and configurations as described in the seminal paper. Reproducing the original work was crucial for validating our modifications and ensuring the reliability of our comparative analysis.

2.4 Exhaust as Many Variation Options as Possible

To thoroughly explore NEAT's potential, we tested a wide range of configuration variations:

- **Initial Number of Nodes:** Starting with different numbers of nodes to assess the impact on learning efficiency.
- **Fitness Function Adaptation:** Customizing the fitness calculation to reward box-pushing and penalize inactivity, aiming to improve convergence rates.
- **Population Size and Generations:** Experimenting with different population sizes and numbers of generations to optimize exploration and solution finding.
- **Mutation Rates:** Introducing random mutations at each step to maintain genetic diversity and prevent premature convergence.
- **Training Environments:** Using multiple random levels to ensure the diversity of training environments and improve generalization.
- **Network Types:** Testing FeedForward = False (RNN) to evaluate the impact of recurrent connections.

These variations were systematically tested to identify the most effective configurations for solving the Sokoban problem.

2.5 Compare and Discuss Results

The final step involved comparing the performance of NEAT against the baseline RL models. Key metrics included the number of iterations required to achieve success, the success rate, and the time per iteration. Our analysis focused on:

- **Customizing Fitness Functions:** Evaluating how tailored fitness functions influenced learning outcomes.
- **Species Diversity:** Assessing the role of species diversity and its impact on maintaining unique network structures and decision-making processes.
- **Network Complexity:** Investigating the effects of different network complexities on problem-solving capabilities.

The results were discussed in detail, highlighting the strengths and limitations of each approach. Our findings underscore the importance of fitness customization, species diversity, and careful management of network complexity in optimizing NEAT's performance for complex tasks like Sokoban. The comparative analysis provided insights into the broader applicability of NEAT and RL algorithms for solving PSPACE-complete problems, contributing to the ongoing research in neuroevolution and reinforcement learning.

3 Simulation Problem Definition

3.1 OpenAI Gymnasium for Zero-Player Games Simulation

OpenAI Gymnasium is a toolkit for developing and comparing reinforcement learning (RL) algorithms. It provides a standardized environment and a diverse collection of tasks (environments) designed to benchmark RL algorithms. Gymnasium is an evolution of the original OpenAI Gym, offering enhancements and a more extensive set of tools to facilitate RL research and development.

OpenAI Gymnasium enables the simulation of zero-player games, where the agent (algorithm) interacts with the environment without human intervention. This setup is crucial for testing and training RL algorithms, as it allows the agent to learn optimal strategies through trial and error within a simulated environment. By providing a consistent interface and a suite of environments, Gymnasium allows researchers to focus on algorithm development and performance evaluation.

OpenAI Gymnasium environments are pre-defined scenarios in which RL agents can be trained and tested. These environments range from simple tasks, such as balancing a pole on a cart (CartPole), to complex simulations like robotic control (Mujoco). Each environment comes with a set of rules and objectives, providing a controlled setting for evaluating the effectiveness of RL algorithms.

In addition to the official environments provided by OpenAI Gymnasium, there are numerous third-party environments created by the community. These third-party environments extend the capabilities of Gymnasium by introducing new challenges and scenarios, allowing for broader testing and application of RL algorithms. Examples include environments for robotics, strategy games, and classic video games.

3.2 The Complexity of Sokoban

Gym-Sokoban is a third-party environment in OpenAI Gymnasium, simulating the classic Sokoban puzzle game. Sokoban is a PSPACE-complete problem where the player (agent) pushes boxes to designated target locations within a grid. The complexity of the game arises from the need for strategic planning and optimal movement sequences to avoid irreversibly blocking boxes. Gym-Sokoban provides a challenging testbed for RL algorithms, requiring them to develop sophisticated strategies to solve the puzzles efficiently.

- **Objective:** Push all boxes to target locations.
- **Environment:** A grid-based puzzle with walls, boxes, and targets.
- **Challenges:** Avoiding deadlocks and planning optimal moves.

The Gym-Sokoban environment is particularly useful for testing the performance of RL algorithms on complex, strategic tasks that require a high degree of foresight and planning.

Sokoban has been extensively analyzed through the lens of computational complexity theory. Initially, it was demonstrated that the computational problem of solving Sokoban puzzles is NP-hard [4,5]. Subsequent research established that it is also PSPACE-complete [6,7].

Computers face significant challenges in solving non-trivial Sokoban puzzles due to the high branching factor, which entails numerous legal moves at each step, and the considerable search depth required, involving many moves to find a solution [8]. Even relatively small puzzles can necessitate extended solutions [9].

Sokoban serves as an excellent testbed for the development and assessment of planning algorithms [10,11]. The first recorded automated solver, Rolling Stone, was created at the University of Alberta. This solver's core methodologies have influenced many subsequent solvers, employing a traditional search algorithm augmented with domain-specific insights [12]. The Festival solver, utilizing the FESS algorithm, was the pioneer in solving all 90 puzzles of the commonly used XSokoban test suite [13]. Nonetheless, even the most advanced automated solvers struggle with many of the more complex puzzles that human solvers can resolve given sufficient time and effort [14].

In computational complexity theory, a decision problem is deemed PSPACE-complete if it can be solved using a memory amount that scales polynomially with the input length (polynomial space) and if every other problem solvable in polynomial space can be polynomially reduced to it. PSPACE-complete problems are considered the most challenging problems within PSPACE, the class of decision problems solvable within polynomial space, because solving any one of these problems would facilitate solving any other problem in PSPACE with relative ease.

Examples of PSPACE-complete problems include determining properties of regular expressions and context-sensitive grammars, verifying the truth of quantified Boolean formulas, identifying step-by-step transitions between solutions of combinatorial optimization problems, and solving various puzzles and games.

4 Baseline Solution - Reinforcement Learning

4.1 Principal Concepts of Reinforcement Learning

Reinforcement Learning (RL) is a type of machine learning where an agent learns to make decisions by taking actions in an environment to maximize cumulative rewards. Key components of RL include the agent, environment, states, actions, rewards, and policy.

- **Agent:** The learner or decision maker.
- **Environment:** Everything the agent interacts with.
- **State** s_t **at time** t: A representation of the current situation of the environment.
- **Action** a_t **at time** t: A set of all possible moves the agent can make.
- **Reward** R_t **at time** t: Feedback from the environment to evaluate the action taken.

110   A. Marchev Jr. et al.

– **Policy π_θ parameterized by θ:** The strategy that the agent employs to determine actions based on the current state.
– **Policy update process $\pi_\theta(s_t) \to \pi_\theta(s_{t+1})$ from state s_t to s_{t+1}:** The implemented change in the policy reflecting the previous states and rewards.

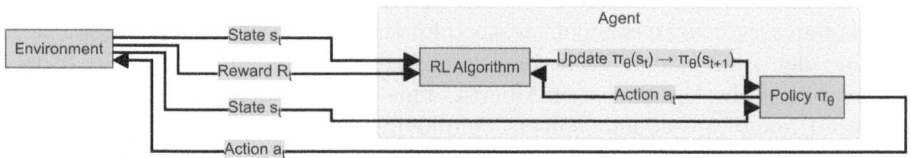

Fig. 1. Flowchart of the principal process of Reinforcement learning

Figure 1 represents a reinforcement learning (RL) process involving an environment and an agent. The environment E provides a reward R_t to the RL algorithm D. The environment E also provides the current state s_t to both the RL algorithm D and the policy π_θ in the agent. The policy π_θ determines an action a_t based on the current state s_t. This action a_t is executed in the environment E, leading to the next state s_{t+1}. The action a_t is also used by the RL algorithm D to update the policy. The RL algorithm D updates the policy π_θ from $\pi_\theta(s_t)$ to $\pi_\theta(s_{t+1})$. This update involves adjusting the parameters θ to improve future actions based on the reward R_t. The updated policy π_θ is used in subsequent interactions with the environment. This loop continues iteratively, refining the policy π_θ to maximize cumulative rewards. The RL algorithm can be any standard algorithm such as Q-learning, Deep Q-Network (DQN), or Proximal Policy Optimization (PPO).

4.2   RL Algorithms Used

Q-Learning is a value-based off-policy RL algorithm that aims to learn the value of the optimal policy independently of the agent's actions. It updates the Q-values (quality of actions) iteratively using the Bellman equation:

$$Q(s,a) \leftarrow Q(s,a) + \alpha[r + \gamma \max_a Q(s',a') - Q(s,a)] \qquad (1)$$

where α is the learning rate, γ is the discount factor, r is the reward, and s', a' are the next state and action.

Deep Q-Network (DQN) extends Q-Learning by using deep neural networks to approximate the Q-values. This allows it to handle high-dimensional state spaces. DQN employs experience replay and target networks to stabilize training.

Proximal Policy Optimization (PPO) [2] is a policy gradient method that aims to balance exploration and exploitation. It optimizes a surrogate objective function while ensuring that the new policy is not too far from the old policy to maintain stable learning:

$$L^{\text{CLIP}}(\theta) = \mathbb{E}_t \left[\min(r_t(\theta)\hat{A}_t, \text{clip}(r_t(\theta), 1-\epsilon, 1+\epsilon)\hat{A}_t) \right] \quad (2)$$

where $r_t(\theta)$ is the probability ratio and $\hat{A}_t$ is the advantage estimate.

PPO with Convolutional Neural Networks (PPO CNN) is a variant of PPO that incorporates CNNs to extract features from high-dimensional input spaces, such as images. This is particularly useful for tasks involving visual data, where CNNs can effectively capture spatial hierarchies in the input.

5 NeuroEvolution of Augmenting Topologies

5.1 Principal Concepts of NEAT

Neuroevolution (NE), the artificial evolution of neural networks using genetic algorithms [1], has been highly effective in reinforcement learning tasks, particularly those with hidden state information. This paper introduces NEAT (NeuroEvolution of Augmenting Topologies), which evolves both neural network topologies and weights. Neuroevolution (NE) has shown promise in reinforcement learning tasks by evolving artificial neural networks (ANNs). However, evolving both topology and weights can enhance performance, although it might complicate the search. NEAT addresses this by:

1. Using historical markings to line up genes for meaningful crossover,
2. Speciating to protect topological innovations,
3. Growing structures incrementally from minimal initial structures.

NEAT enhances neuroevolution by both optimizing and complexifying solutions incrementally. Each component (historical markings, speciation, minimal initial structure) is critical for efficient evolution, making NEAT a powerful approach for complex reinforcement learning tasks.

Genetic Encoding: Each genome in NEAT consists of a list of connection genes and node genes. Connection genes refer to two nodes, specify weights, enable bits, and innovation numbers, which help align genes during crossover. Mutations in NEAT can change connection weights or structures by adding connections or nodes. Figure 2 shows an example of genetic encoding.

Historical Markings: To perform crossover between diverse genomes, NEAT uses historical markings. Each new gene created by mutation is assigned a unique innovation number, which helps identify and match genes from different genomes.

This avoids the problem of competing conventions and simplifies the crossover process.

$$\delta = c_1 \frac{E}{N} + c_2 \frac{D}{N} + c_3 \bar{W} \qquad (3)$$

where:

- E is the number of excess genes,
- D is the number of disjoint genes,
- $\bar{W}$ is the average weight difference of matching genes,
- N is the number of genes in the larger genome,
- c_1, c_2, c_3 are coefficients.

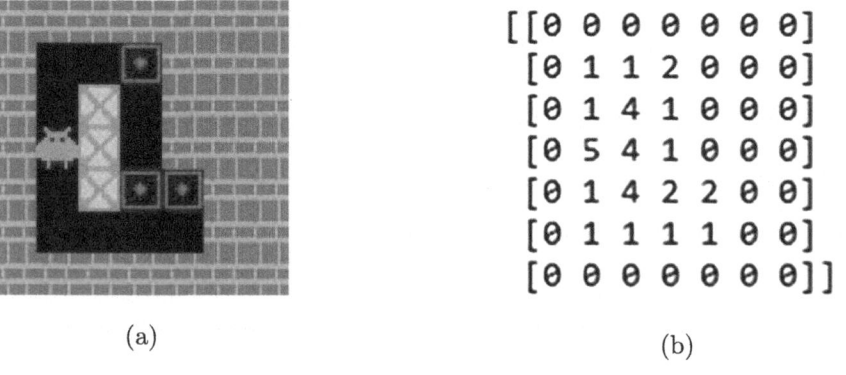

Fig. 2. Example of Genetic encoding - Game level (a), Level Encoding (b), Input Layer Encoding (c)

Protecting Innovation through Speciation: NEAT speciation protects structural innovations by allowing them to optimize without direct competition. Genomes are divided into species based on compatibility distance, ensuring diversity. Fitness sharing adjusts individual fitness by the number of species members, promoting niche preservation (see Fig. 3).

$$f'_i = \frac{f_i}{\sum_{j=1}^{n} sh(d(i,j))} \qquad (4)$$

where:

- f'_i is the adjusted fitness,
- f_i is the original fitness,
- $sh(d(i,j))$ is the sharing function,
- n is the number of individuals.

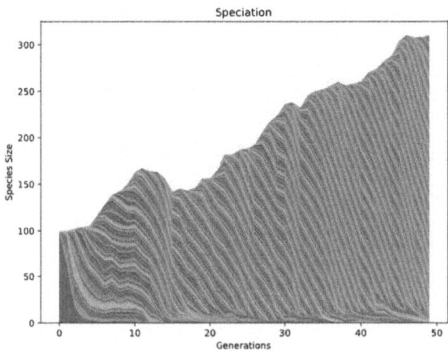

Fig. 3. Increasing Speciation

Minimizing Dimensionality: NEAT starts with a minimal structure (no hidden nodes) and grows only necessary structures through mutations. This reduces the search space dimensions, improving efficiency and avoiding unnecessary complexity.

5.2 Flowchart of NEAT Algorithm

Figure 4 represents the complete flowchart of the NEAT algorithm:

- **Start with Minimal Network Structure:** Initialize each network in the population with the minimal structure, typically consisting only of input and output nodes without any hidden nodes.
- **Initialize Population:** Create an initial population of these minimal networks.
- **Evaluate Fitness:** Assess the performance of each network in the population using a fitness function specific to the task at hand.
- **Termination Check:** Determine if the termination criteria are met (e.g., a network achieves a desired fitness level or a maximum number of generations is reached). If yes, proceed to the best network found.
- **Speciation:** Divide the population into species based on network topological similarities to protect innovation.
- **Selection:** Select the best-performing networks within each species for reproduction based on their fitness.
- **Reproduction:** Generate offspring through reproduction. This involves either mutation or crossover.
- **Mutation Check:** Decide whether a mutation will occur. If yes, proceed to add a connection or node.
- **Add Connection or Node:** Introduce structural mutations by adding a new connection between nodes or inserting a new node into an existing connection.
- **Perform Crossover:** If no mutation occurs, combine genomes from two parent networks to produce offspring.

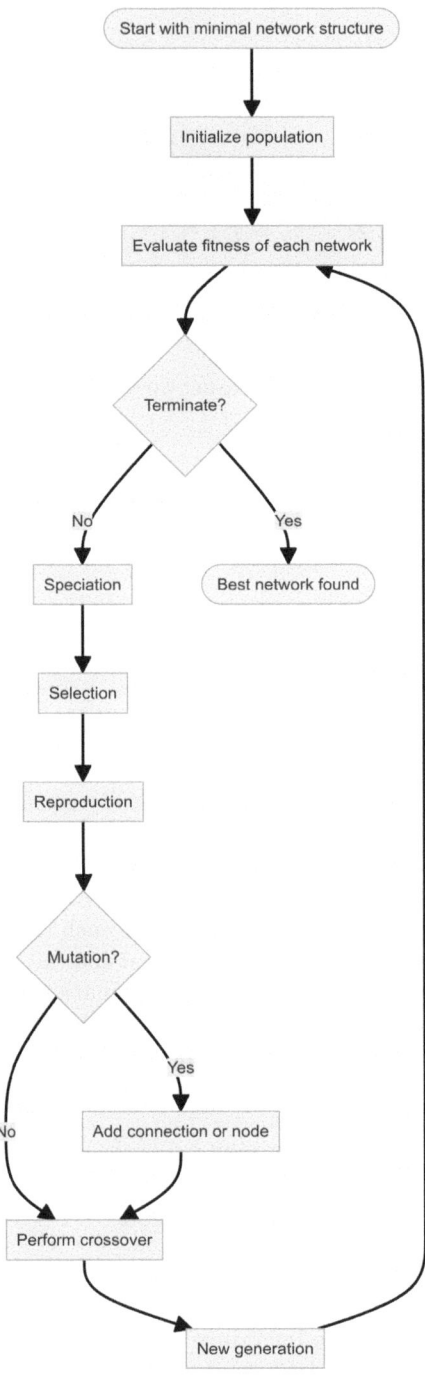

Fig. 4. Flowchart of the principal process of NEAT

- **New Generation:** Form a new generation of networks from the offspring.
- **Repeat Evaluation:** Evaluate the fitness of the new generation and repeat the process from the fitness evaluation step until the termination condition is met.
- **Best Network Found:** Once the termination condition is satisfied, the best network is identified as the result.

6 Implementation of NEAT

For the software implementation the Python NEAT library is used [3]. The main method for modifying the behaviour of the algorithm is through a configuration file for better reproducibility and easier automation.

6.1 Defining the Experiment Using a Configuration

Defining an experiment using the NEAT configuration file involves several key steps, each corresponding to a section of the file. The NEAT algorithm utilizes a variety of parameters to guide the evolution of neural networks. These parameters can be grouped into several categories:

Algorithm Parameters: These parameters define the overall behavior of the NEAT algorithm, such as the fitness criterion used for selection, the threshold at which evolution stops, the population size, and whether the population resets on extinction.

Genome Parameters: These settings govern the default properties of the genomes, including activation functions, aggregation functions, biases, compatibility, connections, nodes, network structure, response, and weights. For activation and aggregation functions, the default function, mutation rate, and available options are specified. Biases parameters relate to initialization, mutation, and replacement. Compatibility parameters include coefficients for disjoint genes and weight differences. Connection and node parameters define probabilities for adding or deleting connections and nodes. Network structure parameters include initial connection settings, feed-forward configuration, and the number of input, hidden, and output nodes. Response parameters specify initialization, mutation, and replacement for node responses. Weight parameters cover initialization, mutation, and replacement for connection weights.

Speciation Parameters: These parameters determine how genomes are grouped into species to maintain diversity within the population. This includes the compatibility threshold for considering genomes as part of the same species.

Stagnation Parameters: These parameters address the conditions under which species are considered stagnant and the criteria for maintaining or eliminating such species. This includes the method for calculating species fitness, the maximum number of generations a species can stagnate, and the number of elite species preserved.

Reproduction Parameters: These parameters govern the reproductive process, including the number of elite individuals preserved each generation and the proportion of individuals that survive to the next generation.

6.2 Experimental Design

Each set of experiments is designed to investigate how different configurations of NEAT parameters impact the evolution of neural networks. The parameters vary in aspects such as connection management, node handling, mutation rates, and species management, all of which play critical roles in shaping the networks' ability to learn and solve tasks. By systematically varying these parameters, the study aims to uncover the most effective configurations for evolving robust

Table 1. Configurations for NEAT experiments

parameter	Experimental set 1			Experimental set 2		
	config-1	config-2	config-3	config-4	config-5	config-6
fitness_threshold	10	10	10	7	7	7
pop_size	1000	500	200	500	500	500
activation_default	sigmoid	sigmoid	relu	sigmoid	sigmoid	sigmoid
activation_mutate_rate	0.05	0.1	0.1	0.03	0.03	0.03
aggregation_mutate_rate	0	0.1	0.1	0	0	0
bias_mutate_power	0.5	1	1	0.88	0.88	0.88
bias_mutate_rate	0.7	0.7	0.7	0.7	0.3	0.7
conn_add_prob	0.5	0.7	0.7	0.7	0.3	0.7
conn_delete_prob	0.5	0.2	0.2	0.4	0.15	0.4
enabled_mutate_rate	0.01	0.05	0.05	0.01	0.01	0.01
feed_forward	1	1	1	1	0	1
node_add_prob	0.3	0.5	0.5	0.5	0.3	0.6
node_delete_prob	0.15	0.2	0.2	0.3	0.15	0.3
num_hidden	1	2	0	0	0	0
response_mutate_power	0	1	1	0.88	0.88	0.88
response_mutate_rate	0	0.7	0.7	0.7	0.7	0.7
response_replace_rate	0	0.1	0.1	0.1	0.3	0.3
weight_mutate_power	0.5	1	1	0.88	0.88	0.88
weight_mutate_rate	0.8	0.8	0.8	0.8	0.4	0.8
weight_replace_rate	0.1	0.1	0.1	0.1	0.1	0.3
compatibility_threshold	3	2	2	2	2.5	2
max_stagnation	20	10	10	10	15	10
species_elitism	2	1	2	2	2	2
elitism	2	1	5	2	2	2

and efficient neural networks. See Table 1 for precise parameter values for each experiment.

Experimental Set 1. In the first set, the experiments share common parameters that set a baseline for comparison. These parameters include the fitness threshold, mutation rates for biases and weights, and the structure of the neural network. These parameters are essential as they determine the basic setup of the neural networks, including how they evolve over generations and adapt to the given task.

Configuration 1 - This configuration uniquely emphasizes higher probabilities for connection deletion and lower mutation rates for enabling connections. It features a minimal hidden layer architecture, with specific settings for node addition and deletion rates, which are crucial for exploring the network's complexity. It also focuses on limited mutation power and rates for response parameters, with a higher compatibility threshold to encourage diversity.

Configuration 2 - Configuration 2 aligns closely with Configuration 3 but includes specific adjustments in the number of hidden nodes and elitism values. These settings are intended to balance the evolutionary pressure by controlling how many top-performing individuals are preserved each generation. The higher node addition probability and lower node deletion rate aim to incrementally increase the network's complexity.

Configuration 3 - Similar to Configuration 2, this configuration modifies the number of hidden nodes and adjusts the elitism values to preserve more top performers. It retains higher mutation power for responses and a moderate compatibility threshold to maintain species diversity, ensuring a robust exploration of possible solutions.

Experimental Set 2. The second set of experiments introduces different common parameters to create a new baseline for the experiments. This includes a lower fitness threshold, a fixed population size, and specific activation and aggregation mutation rates. These parameters ensure that the neural networks are initialized with consistent properties and evolve under controlled mutation rates, focusing on how activation and aggregation functions impact performance.

Configuration 5 - Configuration 5 distinguishes itself with a focus on lower connection deletion probabilities and a non-feed-forward architecture, which allows recurrent connections. This setup aims to explore how recurrent connections can influence learning and adaptation. It also features lower mutation rates for weights and specific settings for node and response mutations to test their impact on the network's adaptability and stability.

Configuration 4 - This configuration shares several parameters with Configuration 6 but stands out with its unique settings for node addition and response replacement rates. It emphasizes higher mutation rates for weights and a standard feed-forward structure, aiming to balance network complexity and mutation stability.

Configuration 6 - Configuration 6 modifies the node addition and response replacement rates to further explore their effects on network evolution. It retains the higher mutation rates for weights and a feed-forward structure, similar to Configuration 4, ensuring a focused comparison on how these parameters influence the evolutionary dynamics.

7 Key Findings and Conclusions

7.1 Solvability of PSPACE-Complete Problem Using NEAT

The NEAT algorithm was tested on the PSPACE-complete problem of solving the Sokoban puzzle. NEAT was able to successfully solve the Sokoban problem within 1000 iterations (20 generations), outperforming traditional reinforcement learning (RL) algorithms such as DQN, PPO, and their variants, underscoring its potential in solving complex PSPACE-complete problems without predefined network architectures. Table 2 presents the empirical results from all experiments, including the baseline models.

Table 2. Empirical results from all experiments

algo	variant	iters	success	time/iter
RL	DQN	5000	No	45
RL	PPO	5000	No	26
RL	PPO opt	5000	No	29
RL	PPO CNN	5000	No	27
RL	Q-learn	1000	Yes	23
NEAT	config 1	1000	Yes	11
NEAT	config 2	1000	No	14
NEAT	config 3	1000	No	13
NEAT	config 4	1000	No	15
NEAT	config 5	1000	No	16
NEAT	config 6	1000	No	17

Different configurations of NEAT were extensively tested to identify important factors of solvability. Configuration 1's success highlights the importance of higher probabilities for connection deletion, lower mutation rates for enabling connections, and maintaining minimal hidden layer architecture for effective problem-solving.

Traditional RL algorithms like DQN, PPO, PPO optimized, and PPO CNN did not achieve success in solving the Sokoban problem within 5000 iterations. Q-learning, a simpler RL variant, successfully solved the problem, suggesting its potential applicability alongside NEAT.

Fitness customization played a crucial role in improving NEAT's learning process, enabling the algorithm to reward strategic moves and penalize inactivity. Species diversity, ensured through NEAT's speciation mechanism, was vital in maintaining unique network structures and promoting innovative solutions. Managing network complexity by starting with minimal structures and allowing incremental growth through mutations proved effective in navigating the solution space efficiently.

7.2 The Configuration Setups

Through various tests with different configuration setups, we found:

- **Customizing Fitness:** Customizing the fitness function improved the learning process.
- **Feedforward vs. Recurrent:** No significant difference was observed between feedforward and recurrent networks (feedforward = False) for Sokoban during the first 50 generations with a population of 500.
- **Species Influence:** The number of species greatly influences behavior:

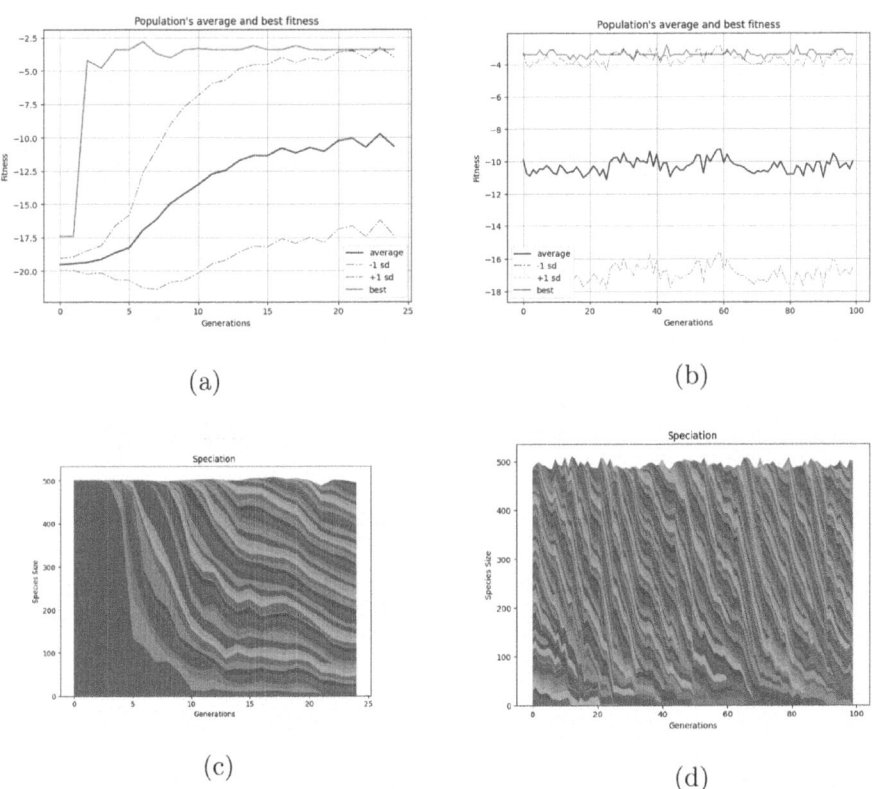

Fig. 5. Training results (a, b) vs. respective speciation (c, d)

- Higher number of hidden nodes (and layers) proportionally increases the number of species.
- More hidden nodes lead to greater diversity and a higher chance of improving fitness scores and finding optimal solutions (Fig. 5).

7.3 The Software Implementation

The neat-python library's mechanisms provide critical insights into the functioning of NEAT:

- **Speciation:**
    - The population is divided into species based on genetic similarity.
    - Speciation protects innovation by preventing newly mutated genomes from competing directly with more mature ones.
    - Each species is assigned a fitness score, typically the average fitness of its members.
    - Speciation allows search to proceed in multiple spaces simultaneously.
    - Without speciation, structural innovations do not survive, and the population quickly converges on initially well-performing topologies.
- **Fitness Sharing:** Reduces the fitness of similar individuals within a species to encourage diversity and prevent any single individual from dominating.
- **Selection:**
    - Parents are selected based on their fitness, with fitter individuals having a higher selection probability.
    - **Stochastic Universal Sampling (SUS):** Ensures a more even distribution of offspring among individuals according to their fitness.
    - Neat-python uses a replacement strategy where a portion of the least fit individuals are replaced by new offspring. If a species stagnates after a certain number of generations, it can be replaced.
- **Crossover:**
    - Neat-python uses a specialized crossover mechanism that aligns genes from both parents to preserve as much functionality as possible.
- **Mutation:**
    - Direct influence on parameters such as weights, biases, number of nodes, and connections.
    - Cannot directly influence actions taken at each step (this was additionally implemented by us).
- Technical limitations: Scaling the solution is limited to CPU optimizations due to the nature of the neat-python and gym-sokoban libraries.

7.4 Future Research

Future goals for this research include exploring custom neuro-evolutionary architecture search, which involves developing different genome structures and initialization methods. Additionally, improving the crossover mechanisms and initial hyperparameter optimization are key areas of focus. These enhancements aim to reduce the necessity of experimenting with different configuration files by optimizing these parameters from the outset. By advancing these aspects, the efficiency and effectiveness of neuro-evolutionary algorithms like NEAT can be significantly improved, facilitating more robust and adaptive solutions to complex problems.

Acknoledgements. This work was financially supported by the UNWE Research Programme.

References

1. Stanley, K.O., Miikkulainen, R.: Efficient evolution of neural network topologies. In: Proceedings of the 2002 Congress on Evolutionary Computation, CEC 2002 (Cat. No.02TH8600), vol. 2, pp. 1757–1762. IEEE, Honolulu (2002). https://doi.org/10.1109/CEC.2002.1004508
2. Schulman, J., Wolski, F., Dhariwal, P., Radford, A., Klimov, O.: Proximal policy optimization algorithms. arXiv preprint arXiv:1707.06347 (2017). https://doi.org/10.48550/arXiv.1707.06347
3. McIntyre, A.: NEAT-Python 0.92 documentation. CodeReclaimers, LLC (2019). https://neat-python.readthedocs.io/. Accessed 1 July 2024
4. Fryers, M., Greene, M.: Sokoban. Eureka **54**, 25–32 (1995)
5. Dor, D., Zwick, U.: SOKOBAN and other motion planning problems. Comput. Geom. **13**(4), 215–228 (1999). https://doi.org/10.1016/S0925-7721(99)00017-6
6. Culberson, J.C.: Sokoban is PSPACE-complete. Technical report TR 97-02, Department of Computing Science, University of Alberta (1997)
7. Hearn, R.A.: Games, puzzles, and computation. Ph.D. thesis, Massachusetts Institute of Technology, pp. 98–100 (2006)
8. Junghanns, A., Schaeffer, J.: Sokoban: improving the search with relevance Cuts. Theoret. Comput. Sci. **252**(1–2), 5–19 (2001). https://doi.org/10.1016/S0304-3975(00)00080-3
9. Holland, D., Shoham, Y.: Theoretical analysis on Picokosmos 17 (2016)
10. Junghanns, A., Schaeffer, J.: Sokoban: evaluating standard single-agent search techniques in the presence of deadlock. Theoret. Comput. Sci. **252**(1–2), 4–19 (1998)
11. Virkkala, T.: Solving sokoban. MSc thesis, University of Helsinki, p. 1 (2011)
12. Junghanns, A., Schaeffer, J.: Sokoban: enhancing general single-agent search methods using domain knowledge. Artif. Intell. **129**(1–2), 219–251 (2001). https://doi.org/10.1016/S0004-3702(01)00109-6
13. Shoham, Y., Schaeffer, J.: The FESS algorithm: a feature based approach to single-agent search. In: 2020 IEEE Conference on Games (CoG), pp. 1–8. IEEE, Osaka (2020). https://doi.org/10.1109/CoG47356.2020.9231929
14. Damgaard, B.: Open test suite - numbers. sokoban solver statistics (2024). https://sokoban-solver-statistics.sourceforge.io/statistics/OpenTestSuite/Numbers.html

Investigating the Regularization of Deep Neural Networks for Affect Recognition with Relevance-Guided Local Explanations

Ines Rieger[1,2]

[1] Chair for Cognitive Systems, University of Bamberg, Bamberg, Germany
ines.rieger@uni-bamberg.de
[2] Fraunhofer Institute for Integrated Circuits IIS, Erlangen, Germany

Abstract. Deep neural networks (DNNs) have demonstrated remarkable performance in various computer vision tasks. However, they face challenges that can inhibit their performance and transparency such as the learning of spurious patterns and a lack of explanatory power. This paper addresses these challenges in the domain of affect recognition, particularly for facial expressions. Our first contribution focuses on the integration of domain-specific knowledge into DNNs. To achieve this, we improve on a regularization method that constrains class co-occurrences, thereby outperforming existing state-of-the-art approaches. Our second contribution evaluates the impact of this regularization by employing an adapted explainable AI (XAI) method that incorporates expert knowledge. The results reveal that the regularization term encourages the learning of more generalized features. Consequently, XAI methods enhance the transparency of DNNs, contributing to the development of more reliable AI systems.

Keywords: Explainable AI · Deep Neural Networks · Regularization

1 Introduction

Deep neural networks (DNNs) have demonstrated exceptional performance in various computer vision tasks. However, they face limitations that inhibit their performance and interpretability, such as learning spurious patterns [10] and not providing explanations for their predictions [11].

In deep learning, *spurious patterns* are associations learned by a model that do not reflect the true underlying structure of the data. These patterns can arise due to data scarcity or low data quality. This paper addresses this challenge for the field of automatic affect recognition. We focus on facial expressions as described by the psychological framework known as the Facial Action Coding System (FACS) [5]. FACS defines modular Action Units (AUs) for distinct facial movements (e.g., outer brow raiser, chin raiser). Combinations of AUs can characterize nearly any facial expression (e.g. happiness, sadness) [6]. Training DNNs to detect AUs can be challenging due to limited data availability, as expert annotators are required. This often results in smaller datasets with limited number and variety of subjects or recording settings, introducing spurious patterns and challenging the generalizability of trained DNNs. To enhance the models' generalizability, integrating domain-specific knowledge as constraints is beneficial [3,9,17]. For the multi-label, multi-class problem of AU detection, co-occurrences that demonstrate the relationships between AU classes can serve as valuable domain knowledge. This information is relatively independent of dataset properties, such as recording settings or subject metadata. Including these constraints can improve model performance and generalizability [9,17]. Additionally, integrating related tasks is also beneficial [16]. For instance, Shao et al. [16] pair AU detection with optical flow estimation, while Yang et al. [17] propose embedding semantic AU descriptions in a multi-modal approach. Rieger et al. [9] exploit co-occurrences as constraints in the loss function (CorrLoss). This paper explores and improves on the lightweight and flexible CorrLoss method [9]. Our first research contribution is as follows: *We propose to adapt CorrLoss [9] by using a different distance metric. The adapted CorrLoss demonstrates superior performance, also compared to other state-of-the-art approaches.*

However, performance metrics alone may not provide a comprehensive understanding of the internal workings of trained DNNs. AI systems, such as DNNs, are often *opaque black boxes*, making it crucial to understand their behavior to develop trustworthy AI systems [12]. As AI applications become increasingly central to our lives, the need for transparency and explainability grows. To address this, explainable artificial intelligence (XAI) approaches are pivotal. We propose applying XAI methods to evaluate the impact of the CorrLoss regularization method on DNN features, thereby increasing the transparency of the underlying changes. Local explanation methods, such as gradient-weighted class activation mapping (Grad-CAM) [13] and layer-wise relevance propagation (LRP) [1], are commonly used to visualize DNN decisions. These methods produce heatmaps highlighting the relevant areas of an image influencing the model's decision. However, examining heatmaps for a large number of images is impractical. To address this, Rieger et al. [8] propose a verification pipeline that quantifies heatmaps of AU detection models based on relevant areas from expert knowledge. We adapt their approach (see Fig. 2) to investigate the impact of regularizing a DNN with CorrLoss. Our second research contribution is as follows: *We apply relevance-guided local explanations to evaluate how the features change when a DNN for affect recognition is regularized with CorrLoss. The results demonstrate that using CorrLoss yields more generalized features across classes, contributing to performance gains.*

This work aims to advance the development of evaluation methods that shed light on the inner workings of DNNs and thus facilitate the development of trustworthy AI systems [12]. By increasing transparency, we can ensure that AI systems are not only powerful, but also reliable and understandable.

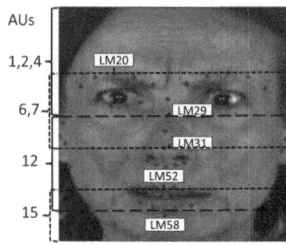 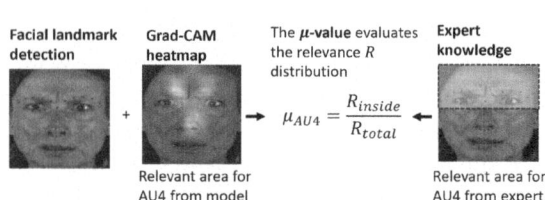

Fig. 1. Overview of expert knowledge [5] illustrating the relevant areas for AU classes. Coordinates from facial landmarks (LMs).

Fig. 2. Relevance-guided local explanations use expert knowledge to quantify heatmaps, providing a detailed understanding of the inner workings of a trained DNN. Image is from Actor Study [14].

2 Methods

2.1 Regularization Term CorrLoss

This paper adapts the CorrLoss regularization term from [9]. CorrLoss is a weighted regularization term designed to integrate the co-occurrence between classes. The formulation from [9] is generalized in Eq. 1 to represent any p-norm as distance metric.

$$R_{corr}(AU) = \frac{\left(\sum_{1 \leq j \leq i \leq N_{AU}} |c(y_i, y_j) - c(\hat{y}_i, \hat{y}_j)|^p\right)^{\frac{1}{p}}}{B} \quad (1)$$

The variable c denotes the statistical relationship between AU classes (N_{AU}) in an $n \times n$ matrix. Only unique pairs below the diagonal are used. The relationship strength is computed using the Pearson Correlation Coefficient (PCC). During training, the aim is to minimize the difference between the correlation strength of the ground truth labels $c(y_i, y_j)$ and the predicted labels $c(\hat{y}_i, \hat{y}_j)$. To calculate this distance, we propose to compare two different p-norms: the Manhattan norm (L_1) as used in [9], and the Euclidean norm (L_2). The aggregated distances are normalized by dividing by the count of unique class correlations B. R_{corr} is averaged across the entire training batch. The final loss is a weighted combination of the binary cross-entropy loss and the regularization term R_{corr}. The best weight for each model is determined from the range [0, 1] with a step size of 0.1.

2.2 Evaluate DNNs Using Relevance-Guided Local Explanations

Figure 2 provides an overview of our relevance-guided local explanation method, based on [8], to verify the impact of the regularization method CorrLoss beyond performance metrics. This approach uses bounding boxes defined from expert knowledge to quantify the heatmaps, with relevant areas defined by facial landmark coordinates. For landmark detection, we use OpenCV [2]. Using FACS [5], we extend Rieger et al.'s [8] bounding boxes to align with our AU classes (see Fig. 1). To generate heatmaps, we opt for Grad-CAM [13] over LRP [1] due to its lower dependency on hyperparameters, despite being less fine-grained. Future work could compare different local explanation methods to balance granularity and usability. As a metric, we use the μ-value to evaluate if the heatmap relevance distribution matches domain knowledge. The μ-value is calculated as $\mu = \frac{R_{inside}}{R_{total}}$ [8], representing the proportion of relevance within the bounding box (R_{inside}) relative to the entire image (R_{total}). Lower μ-values indicate more dispersed relevance, extending beyond class bounding boxes.

2.3 Experimental Setup

This paper utilizes the Actor Study [14] and BP4D-Spontaneous [18] databases, both annotated with AUs. The Actor Study database comprises 21 subjects and 777 video sequences recorded in a controlled lab setting, focusing on emotions and AUs. It should be noted that this dataset is not fully openly accessible. The BP4D-Spontaneous dataset features 328 videos from 41 subjects, who were stimulated to display eight target emotions. To ensure effective training, we select frames from AU classes that occur in emotions [6] and exceed specific thresholds (3,000 frames for Actor Study; 7,000 frames for BP4D). This selection results in a total of 12,705 frames for the Actor Study and 40,533 frames for BP4D.

We utilize a 50-layer ResNet [7] architecture as backbone, pretrained on ImageNet [4]. Following the average pooling layer, we customize the architecture by adding one dense layer (dim. = 256) using dropout (prob. = 0.5) and an output layer with a sigmoid activation function. We determine the best batch size (Actor Study: 128, BP4D: 64) and use the AMSGrad optimizer (lr = 10^{-4}).

3 Results

Table 1 presents the results of training our DNN with no regularization (None) versus training it using the regularization term CorrLoss with different distance metrics. We evaluate our models using the performance metric F1-score and the μ-value. For the μ-value, as with [8], we only consider correctly predicted instances that are labeled as active AU classes. The μ-value is expressed as a percentage, indicating the proportion of relevance located inside the respective bounding box (e.g., 71.5%). All results are averaged over a 3-fold cross-validation. Based on the average F1-score, our introduced $L2$ norm outperforms the original $L1$ norm [9] for both datasets. The μ-values exhibit considerable variation between classes, with the two classes AU6 and AU7 showing the lowest μ-values. Regarding the effect of regularization, we observe that while the average performance is highest for CorrLoss using the $L2$-norm, the average μ-value is the lowest. We hypothesize that training with the CorrLoss and $L2$-norm likely enforces the strongest class correlations compared to the other configurations. This enforcement may reduce the presence of class-specific features, as indicated by more dispersed relevance and lower μ-values, leading to the development of more generalized features that enhance overall model generalization.

Table 1. Evaluation of DNNs for AU detection with a performance metric (F1-score) and relevance-guided local explanations (μ-value). Best results are bold.

Dataset	Metric	Reg.	AU1	AU2	AU4	AU6	AU7	AU12	AU15	Av.
Actor Study	F1-score	None	**85.2**	69.0	85.8	81.2	62.1	86.9	–	78.4
		$L1$	85.1	68.0	**85.9**	79.6	**71.5**	82.3	–	78.7
		$L2$	85.0	**70.1**	85.7	**81.8**	63.8	**90.0**	–	**79.4**
	μ-value	None	**71.5**	**76.5**	49.7	31.7	37.3	54.1	–	**53.5**
		$L1$	69.4	73.4	43.9	**33.2**	**40.9**	**55.2**	–	52.7
		$L2$	71.0	72.8	**50.4**	30.2	33.5	51.8	–	51.6
BP4D	F1-score	None	58.2	51.8	60.5	**83.0**	86.8	90.5	38.1	67.0
		$L1$	61.9	**44.9**	59.7	81.1	85.1	90.7	41.2	66.4
		$L2$	**62.3**	44.7	**63.1**	81.4	**87.2**	**92.1**	**47.0**	**68.3**
	μ-value	None	61.5	**71.1**	52.4	24.1	34.3	56.7	55.2	50.8
		$L1$	**66.1**	63.2	**59.8**	24.5	32.1	**59.7**	**61.8**	**52.5**
		$L2$	62.6	64.9	55.5	**25.4**	**37.5**	53.2	54.7	50.5

Table 2 presents a comparison between our best adapted CorrLoss, the original CorrLoss [9], and other state-of-the-art approaches. It should be noted that the result for CorrLoss [9] does not include AU4 and 15. Despite this limitation, the results demonstrate that our adapted CorrLoss achieves superior performance compared to the other approaches across the two applied datasets.

Table 2. State-of-the-art comparison. Results are in F1-score. Best are bold.

Dataset	JAO [16]	SEV-Net [17]	CorrLoss [9]	AUReader [15]	CorrLoss(our)
Actor Study	–	–	–	73.2	**79.4**
BP4D	64.7	66.1	63.8	–	**68.3**

4 Conclusion

This paper addresses the challenges of deep learning in affect recognition and proposes a solution by integrating domain-specific knowledge into DNNs and utilizing explainable AI (XAI) techniques for evaluation. Our approach enhances the state-of-the-art performance by adapting the regularization method CorrLoss, which constrains class co-occurrences. Additionally, we evaluate the effect of CorrLoss on DNN features by adapting a relevance-guided local explanation method. Our findings indicate that CorrLoss improves the performance of the model by promoting the learning of generalized features. By understanding the underlying mechanisms of DNNs, this research aims to advance the development of trustworthy AI systems.

References

1. Bach, S., Binder, A., Montavon, G., Klauschen, F., Müller, K.R., Samek, W.: On pixel-wise explanations for non-linear classifier decisions by layer-wise relevance propagation. PloS One (2015)
2. Bradski, G.: The openCV library. Dr. Dobb's J.: Softw. Tools Prof. Program. (2000)
3. Dash, T., Chitlangia, S., Ahuja, A., Srinivasan, A.: A review of some techniques for inclusion of domain-knowledge into deep neural networks. Sci. Rep. (2022)
4. Deng, J., Dong, W., Socher, R., Li, L.J., Li, K., Fei-Fei, L.: ImageNet: a large-scale hierarchical image database. In: Proceedings of CVPR. IEEE (2009)
5. Ekman, P., Friesen, W.V.: Facial Action Coding Systems. Environ. Psychol. Nonverbal Behav. (1978)
6. Friesen, W.V., Ekman, P.: EMFACS-7: emotional facial action coding system, version 7. Unpublished manuscript (1983)
7. He, K., Zhang, X., Ren, S., Sun, J.: Deep residual learning for image recognition. In: Proceedings of CVPR (2016)
8. Rieger, I., Kollmann, R., Finzel, B., Seuß, D., Schmid, U.: Verifying deep learning-based decisions for facial expression recognition. In: Proceedings of ESANN (2020)
9. Rieger, I., Pahl, J., Finzel, B., Schmid, U.: CorrLoss: integrating co-occurrence domain knowledge for affect recognition. In: Proceedings of ICPR. IEEE (2022)
10. Sagawa, S., Raghunathan, A., Koh, P.W., Liang, P.: An investigation of why overparameterization exacerbates spurious correlations. In: Proceedings of ICML (2020)
11. Samek, W., Montavon, G., Lapuschkin, S., Anders, C.J., Müller, K.R.: Explaining deep neural networks and beyond: a review of methods and applications. In: Proceedings of IEEE (2021)
12. Schmid, U.: Trustworthy artificial intelligence: comprehensible, transparent and correctable. In: Introduction to Digital Humanism: A Textbook. Springer (2023)
13. Selvaraju, R.R., Cogswell, M., Das, A., Vedantam, R., Parikh, D., Batra, D.: Grad-CAM: visual explanations from deep networks via gradient-based localization. In: Proceedings of ICCV (2017)
14. Seuss, D., et al.: Emotion expression from different angles: a video database for facial expressions of actors shot by a camera array. In: Proceedings of ACII (2019)
15. Seuss, D., et al.: Automatic estimation of action unit intensities and inference of emotional appraisals. IEEE Trans. Affect. Comput. **14**(2), 1188–1200 (2023)
16. Shao, Z., Zhou, Y., Li, F., Zhu, H., Liu, B.: Joint facial action unit recognition and self-supervised optical flow estimation. Pattern Recogn. Lett. (2024)
17. Yang, H., Yin, L., Zhou, Y., Gu, J.: Exploiting semantic embedding and visual feature for facial action unit detection. In: CVPR (2021)
18. Zhang, X., et al.: BP4D-spontaneous: a high-resolution spontaneous 3D dynamic facial expression database. Image Vision Comput. (2014)

Layered Data-Centric AI to Streamline Data Quality Practices for Enhanced Automation

Muhammad Uzair Akmal[1](✉)[iD], Saara Asif[1][iD], Leonid Koval[1][iD], Selvine G. Mathias[1][iD], Simon Knollmeyer[1][iD], and Daniel Grossmann[2][iD]

[1] AImotion Bavaria, Technische Hochschule Ingolstadt, 85049 Ingolstadt, Germany
{MuhammadUzair.Akmal,Saara.Asif,Leonid.Koval,SelvineGeorge.Mathias, Simon.Knollmeyer}@thi.de

[2] Faculty of Computer Science and Data Processing, Technische Hochschule Ingolstadt, 85049 Ingolstadt, Germany
Daniel.Grossmann@thi.de

Abstract. Most artificial intelligence (AI) applications are designed under the model-centric AI (MCAI) approach, where data scientists aim to optimize the machine learning (ML) models starting with fixed, pre-processed data. However, businesses often struggle with limited datasets, changes in data over time, and limited ML knowledge, making it difficult to maintain data quality. One potential approach is data-centric AI (DCAI), which systematically improves the data quality used to build AI systems. However, rapid growth in data volume leads to challenges in selecting the most suitable operations for enhancing data quality thereby maintaining data accuracy, completeness, consistency, and reliability in real-world applications. To address this gap, we propose a novel framework, namely Layered Data-Centric AI (LDCAI) by expanding upon the existing DCAI Pipeline. LDCAI employs a three-layered approach, starting with the DCAI layer which generates the initial training data. This data is then analyzed and refined in the Data Analysis (DA) layer by a team of data scientists and domain experts. The refined data is later sent to the Data Quality Control (DQC) layer for additional quality checks and improvements by data scientists. The entire process can be iterated as needed to continuously improve data quality, which can be fed back into the DCAI layer's model to achieve better predictions and outcomes. The implementation is illustrated through a scalable and customizable architecture designed for optimizing data quality within AI systems.

Keywords: Artificial intelligence · Data-centric · Data analysis · Data quality

1 Introduction

Industries and businesses have been creating use cases for automation combining the Internet of Things (IoT) and AI [1] which has resulted in a massive amount of data (text, image, and video). Advanced machine learning methods, like deep learning (DL) [2], are integral for efficiently analyzing and processing the ever-increasing amount of unstructured data generated from multiple sources [3]. AI has always been associated with complicated algorithms and programming, however, in today's machine-learning world, data is the focal point on which everything is built. Still, it is often neglected and not handled aptly in AI projects, and this impacts the accuracy of the ML model, which tends to be lower than expected because of tuning a model built on low-quality data [4].

In real-world domain applications, models learn from data but often face difficulties due to missing data [5], biased data [6], and class imbalance [7] that deteriorates performance [4]. Presently, there are two views on improving AI systems' performance in the ML community: model-centric AI (MCAI) and data-centric AI (DCAI) [8–10]. The performance in MCAI [9] increases by fine-tuning and optimizing the algorithms while keeping the data fixed. In contrast, DCAI focuses on improving data quality and consistency. The parameter optimization approach for tuning and improving the accuracy of ML models can only play its role to a limited extent [11]. On the contrary, actual improvement is needed in the data quality [12]. DCAI shifts the emphasis from better models to better data [9]. It emphasizes that even basic models can perform much better when accessing good-quality, well-annotated, and relevant data [8]. In this research, we briefly present the limitations of MCAI and highlight the need to adapt DCAI considering its benefits on the AI development life-cycle. The primary contributions of this study are summarized as follows:

- We introduce a conceptual framework (LDCAI) that expands the state-of-the-art DCAI pipeline into three distinct layers. We propose analysis and quality principles within the layers for iteratively analyzing and handling data quality to enhance the AI solutions' performance and efficiency.
- We provide a detailed architecture for implementing the proposed framework in real-world AI applications. We have outlined the specific observations of the proposed framework. Moreover, it is ensured that the proposed framework is aligned with the basic concept of DCAI.

The remainder of the paper is organized as follows: we begin with Sect. 2 by briefly explaining the difference between MCAI and DCAI approaches. Section 3 covers the preliminary research of the state-of-the-art approaches for DCAI. We are introducing a conceptual framework in Sect. 4, that expands the DCAI pipeline into a Layered Data-Centric AI (LDCAI) pipeline. Also in this section, we are presenting an implementation architecture for the LDCAI framework. In Sect. 5, we provide a concise overview of the benefits associated with embracing a data-centric AI (DCAI) approach, both from an industrial standpoint and a societal perspective. Finally in Sect. 6, we conclude our paper and provide future work and experimentation directions for the proposed concept.

2 MCAI vs DCAI: Key Contrasts

The fundamental components of all AI systems are data and models, which work together to achieve the desired outcomes. Rather than solely focusing on optimizing model architectures and tuning hyperparameters, the DCAI approach shifts the focus toward enhancing the quality and preparation of the data used for training models [9]. High-quality, well-curated data [13] will lead to more accurate and reliable model outcomes resulting in greater performance improvements than iterative model tuning alone [4,9,12]. Presently, AI solutions are developed using three main approaches: MCAI, DCAI, and a hybrid approach combining both [8]. Thus, it is crucial to understand the key concepts and differences behind these approaches.

2.1 Model-Centric AI

In the MCAI approach, optimizing the algorithm and underlying architecture that make up the AI model is of utmost importance [8]. MCAI pipeline is shown in Fig. 1, where the dataset is kept static after standardized preprocessing and the performance of the AI is improved by iteratively improving the model [8,9,14] through hyperparameter tunning [11] until you get the desired results.

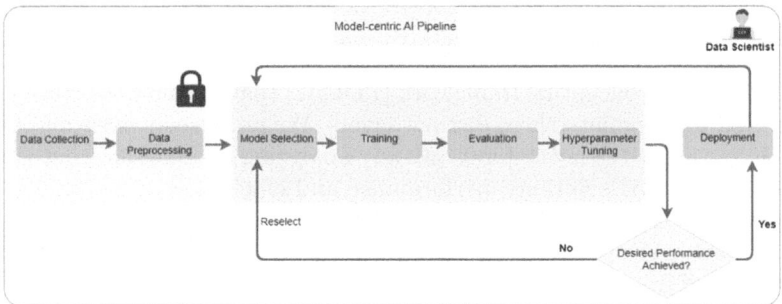

Fig. 1. A typical life cycle of MCAI.

Limitations of Model-Centric AI: MCAI is well-suited for businesses and industries that require generalized solutions [10]. However, standardized AI solutions may not apply to industries like manufacturing, agriculture, and healthcare that require customized AI solutions [14]. These industries should analyze their approach to ensure their model learns from comprehensive and consistently labeled data that covers all important cases [15]. While a model might perform well on the training data, its ability to generalize to new, unfamiliar data can be limited, reducing its applicability to real-world applications. The model may struggle to provide consistent results on data instances it has not encountered during training [10,14]. Some of the major limitations that MCAI suffers from are:

- Reliance on Labeled Data: A large amount of labeled data is required for training models in MCAI [10,14]. However, preparing such large datasets can be time-consuming, costly, and often impractical, particularly in domains (e.g. detecting anomalous activity in video surveillance [16]) with limited availability of labeled data [15].
- Challenges in Model Generalization: Noisy, unclean, and poorly labeled data results in model over-fitting and misinterpretation [4,12]. Models trained on specific datasets may not perform well when applied to different data distributions or situations [14].
- Unclear and Ambiguous Predictions: Complex models when fed with poor-quality data provide hard-to-understand results and make decision-making challenging [4,15].
- High Computational Costs: Labeling large datasets and training models on poor-quality data [4,12] is time-consuming and costly [15,17].

2.2 Data-Centric AI

The DCAI approach focuses on the challenges related to the completeness, consistency, relevance, and diversity of the data used for training and testing the ML models [18]. In the DCAI approach, the focus is shifted from model to data [9]. The idea is to iteratively improve the quality of data [8,9] by keeping the model fixed as shown in Fig. 2. These iterations span from initial training to production deployment and the model is retrained and refined, based on the improved data quality [18,19].

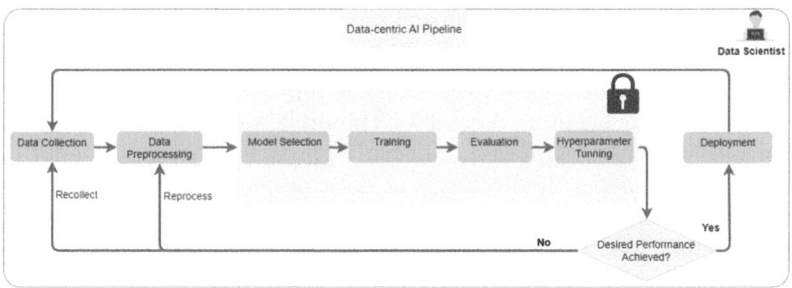

Fig. 2. A typical life cycle of DCAI.

Benefits of Data-Centric AI: Data is the primary focus of the DCAI approach, with significant effort dedicated to understanding and refining it [18,20,21]. By prioritizing data, this approach aims to provide more informed and relevant decisions and results from AI systems. The DCAI approach ensures that data is comprehensive, encompassing all relevant scenarios, while maintaining coherence and being free from contradictions [18]. It aims to tackle the heterogeneity of data sources and representations. This data-driven approach allows for greater flexibility and scalability [19,20,22,23], making it well-suited for handling the ever-growing amount of data and diverse use cases in various domains [18]. Some of the major advantages of DCAI are:

- Improved Model Accuracy: Noise and inconsistencies are significantly reduced from the data, leading to more accurate predictions and model performance [18,19].
- Reduced Bias: Data bias can be identified and mitigated from the dataset by examining and curating the training data that ensures the AI systems are fair and free from discriminatory outcomes [4,12,15].
- Enhanced Interpretability: Good data makes it easier to interpret the results which ensures increased transparency and trust in AI systems [24].
- Better Generalization: Based on high-quality and diversified data, models can generalize better to new and unseen data, enhancing their applicability across different domains [13].
- Faster Development Time: Models fed with clean data require less training time to achieve desired outcomes [14].
- Cost-Effectiveness: Cost overruns due to poor model performance resulting from misrepresentations and erroneous decisions can be avoided by improving data quality [17].
- Encourages Continuous Improvement: A data-centric approach encourages ongoing refinement of data, ensuring that models remain up-to-date and relevant as new data is incurred [18].

3 Related Work

The DCAI paradigm represents a shift in focus from strictly optimizing AI models to prioritizing the quality and preparation of the data used for training. Numerous research efforts have explored this approach, recognizing that better data can lead to more accurate AI results. Hamid [8] explores the constraints and limitations of MCAI and argues a universal approach is not suitable for diverse domains and emphasizes the necessity of data preparation before model training. It advocates for a DCAI approach that focuses on iterative development for data preparation, incorporating successive improvements in data by accessing novel instances thereby enhancing the model's capability to customize solutions effectively.

One of the pioneering works in this field is the paper by Luley [22], in which the author proposes a collaborative approach between the domain expert and the ML engineer where the engineer refines the ML model alongside doing an error analysis of the data for the domain expert to refine and redefine data annotation. It also suggests appropriate tools for the annotation process, communication, and error analysis for the collaboration between the domain expert and the ML engineer. Building upon this foundation, Zha [20] elaborates on the major activities encompassing DCAI as a preliminary step to iterative model training. The work emphasizes the importance of viewing data preparation as more than just a black box, and instead, considering the trade-offs involved, such as data reduction or augmentation, based on feedback from the iterative process. The work proposes to address training and inference data separately. The authors advocate for a comprehensive understanding of the domain and data before deciding whether to use in-distribution or out-of-distribution inference datasets. Data quality assurance is recommended based on feedback from users, domain experts, and stakeholders or algorithmic error analysis.

Recognizing the need for benchmarking and evaluation, Eyuboglu [25] proposes a benchmark for three tasks focused on training data, inference data, and dataset size. The benchmark tasks focus on data cleaning, slice discovery, and dataset reduction to optimize model performance. From a business and information systems engineering perspective, Jakubik [19] introduces a framework to highlight the various aspects of DCAI. The proposed two-step framework addresses dataset refinement and extension based on model evaluation, offering multiple approaches and suggesting individual, organizational, and cross-organizational perspectives to achieve respective objectives. Addressing the concern of energy consumption by AI systems, Verdecchia [26] presents a standpoint of DCAI focused on decreasing the size and improving the data quality, ensuring no loss of accuracy while reducing energy utilization and promoting greener AI.

Kumar [23] outlines a data-centric loss function that minimizes errors in the training data instead of increasing model complexity. The work proposes a general rule that can be adapted to the domain and problem being studied, potentially enabling simpler models to generalize well as the training data undergoes an optimization process for robust feature selection. Recognizing the importance of DCAI for robust and accurate scientific data, Maskey [27] explores this concept from a scientific perspective, while Mazumder [28] presents a benchmarking suite, DataPerf, encompassing data-centric tasks across various modalities, and it is another contribution by expanding the 'just' model-specific study to 'models and datasets'. Finally, Whang [29] looks at the challenges within data-centric AI tasks and proposes a structured pipeline analyzing the sequential and parallel steps involved in collecting, validating, cleaning, and iteratively preprocessing the data.

Data-centric AI has become a crucial concept for improving ML systems in practice [21,24,30]. While existing studies offer valuable insights into DCAI, there remains a significant gap in the effective implementation, primarily due to the absence of standardized processes and key design considerations. This lack of structure makes it challenging for practitioners to adopt and apply DCAI in their work. Addressing this gap is essential for unlocking the full potential of DCAI, thereby enhancing the performance and reliability of AI systems across diverse domains.

4 LDCAI: Layered Data-Centric AI Framework

While the state-of-the-art DCAI pipeline provides abstract insights on enhancing data quality [10,31,32], AI experts lack practical guidance to define and implement the necessary data quality control checks and processes. Consequently, practitioners must often reinvent the entire quality control pipeline from scratch. To address this gap, the Layered Data-Centric AI (LDCAI) framework as shown in Fig. 3, expands the DCAI pipeline by structuring it into three distinct layers:

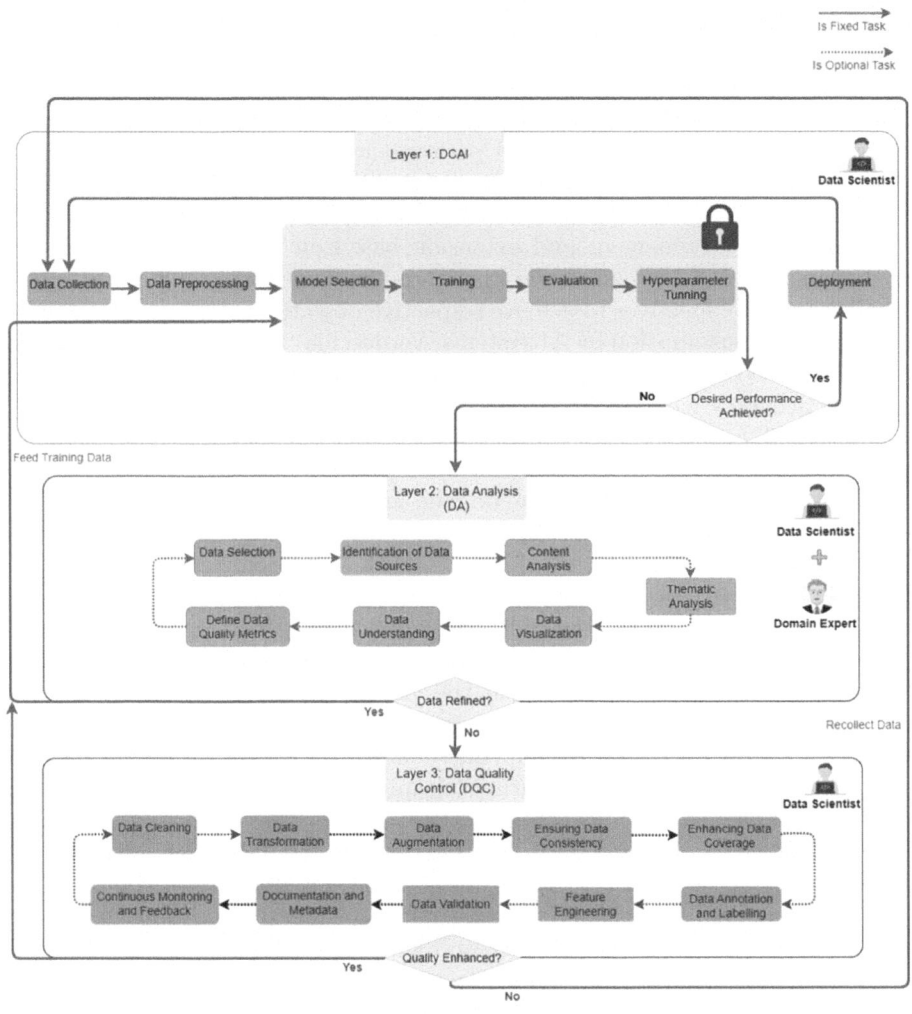

Fig. 3. A typical life cycle of Layered Data-centric AI framework.

Data-Centric AI (DCAI) Layer: This layer encompasses the complete data-centric pipeline starting from data collection till hyperparameter tunning resulting in a trained model. This model can be then either deployed in production or based on the model's poor performance will be passed to the data analysis (DA) layer before deployment.

Data Analysis (DA) Layer: This layer refines the data collection task from the DCAI pipeline by integrating seven principles of data analysis. These principles are not rigidly fixed but can be selected and applied in any order based on the available resources and strategic decisions of the pipeline experts. A comprehensive analysis is performed collaboratively by data scientists and domain experts to ensure effective data examination.

Data Quality Control (DQC) Layer: This layer extends the data preprocessing tasks from the DCAI pipeline, incorporating ten principles for data quality control. Similar to the DA layer, the application of these quality principles is at the discretion of the pipeline experts as dictated by the resources. The purpose of this layer is to ensure rigorous data validation and refinement by data scientists, to address issues such as inconsistencies, outliers, and missing values.

Iterative Quality Checks: After processing in each layer, data scientists perform quality checks to determine if further enhancements are needed. Based on these evaluations, data is either further refined and passed to the subsequent layer or passed to the DCAI layer for model training. This iterative process ensures that data quality is progressively improved before being used for model training.

Model Training and Deployment: Once the data achieves the required quality standards, it is utilized to train the ML model. Upon reaching the desired performance metrics, the trained model is deployed.

Continuous Quality Checks Post-deployment: The framework supports ongoing quality checks post-deployment as well. This continuous monitoring enables identifying and resolving issues such as data drift [33] and concept drift [34], which can impact model performance over time.

By integrating these layers, the LDCAI framework enhances data quality at each stage and establishes a systematic and repeatable process for data refinement. This structured approach mitigates the need for AI practitioners to recreate quality check procedures, thereby fostering efficiency and consistency in the development of robust AI systems. Moreover, the LDCAI framework introduces a flexible approach to implementing data-centric AI, bridging the gap between abstract concepts and real-world implementation. Furthermore, an explanation of all the tasks involved in the LDCAI layers is provided in Table 1.

4.1 Implementation Strategy for LDCAI Framework

The implementation architecture of the LDCAI framework as shown in Fig. 4, is based on microservice architecture [46], where each layer of LDCAI is deployed as a separate microservice. The data scientist is responsible for creating the DCAI pipeline in Node 1. Initially, raw data is ingested at Node 1, which processes it through the DCAI layer, generating the first version of the training data. The first version of the training data is then stored in the data storage component for versioning and forwarded to Node 2 for refinement in the Data Analysis (DA)

Table 1. A detailed task outline for each layer of the LDCAI framework.

Layer	Actors	Task	Explanation and Techniques
Data-centric AI (DCAI) Layer	Data Scientist	Data Collection	Identification of the data sources and acquiring data from different data sources
		Data Preprocessing	Ensuring the correctness and consistency of labels, error analysis, and outliers detection and handling [35]
		Model Selection	Select the models based on the problem type and data attributes. For example, logistic regression, decision trees, random forests, gradient boosting machines, support vector machines, neural networks, etc. can be considered for a classification task [10]
		Training	Train the selected models on the training data
		Evaluation	Evaluate the trained model on the validation set using applicable evaluation metrics (e.g., accuracy, precision, recall, F1-score, etc.) [36]
		Hyperparameter Tuning	Perform hyperparameter optimization for the selected models using techniques like grid search, random search, or Bayesian optimization to find the best configuration for each model [11].
		Deployment	Once the model meets the desired performance criteria, deploy it in a production environment
Data Analysis (DA) Layer	Data Scientist, Domain Expert	Data Selection	Determine the type (e.g. text, image, video, etc.), format (e.g. structured, unstructured, big data, etc.), and amount of data required
		Identification of Data Sources	Identify the sources of data (e.g. databases, public datasets, social media, sensors, user-generated content, etc.)
		Content Analysis	Content analysis can be applied to both qualitative and quantitative data. It focuses on identifying frequencies, recurring words, and subjects from the data [37]
		Thematic Analysis	Thematic analysis can only be applied to qualitative data. It focuses on identifying patterns and themes from the data [37]
		Data Visualization	Visualizations are used to identify patterns, trends, and relationships within the data, which can aid in assessing data quality. Techniques like scatter plots, histograms, and box plots can be used to identify outliers, skewed distributions, and anomalies that may indicate data quality issues [38]
		Data Understanding	Through domain expertise identify any assumptions and constraints related to the data. Use domain knowledge to interpret the data correctly, and understand the importance of various variables and their expected behavior. Understand the data relationships like correlation analysis and cross tabulation [39]
		Define Data Quality Metrics	Establish measurable criteria to determine if the data meets the desired quality standards by defining metrics for dimensions like completeness, validity, timeliness, and consistency [40,41]
Data Quality Control (DQC) Layer	Data Scientist	Data Cleaning	Fix data entry errors, inconsistencies, and anomalies, such as incorrect entries, outliers, and miss-classifications. Identify and remove duplicate records to prevent bias and redundancy. Handle missing values in the data using techniques like mean/mode imputation, regression, or more advanced methods like k-nearest neighbors (KNN) [5,42]
		Data Transformation	Apply normalization and standardization by scaling the numerical data to a standard range or distribution to ensure consistency. Encode categorical variables by converting categorical data into numerical formats using techniques like one-hot encoding or label encoding [43]
		Data Augmentation	Create synthetic data to increase the size of dataset and its diversity especially in cases when limited real-world data is available. For image data, use techniques like rotation, scaling, and flipping to create more training samples [44]
		Ensuring Data Consistency	Establish and apply standards for data entry to minimize variations and errors. Ensure that the data schema (structure, types, and formats) remains consistent across the dataset
		Enhancing Data Coverage	Collect more samples from diverse sources to increase the volume of data. Fix class imbalance issue by collecting more data for underrepresented classes or using techniques like oversampling and undersampling [7]
		Data Annotation and Labelling	Verify that data is accurately labeled with the help of domain experts. Use crowd-sourcing for large-scale labeling tasks and apply verification steps to ensure accuracy [15]
		Feature Engineering	Identify new features that capture important patterns and relationships in the data. Remove irrelevant or redundant features to reduce noise and improve model performance by using techniques like dimensionality reduction and sampling [45]
		Data Validation	Establish and apply automated checks to validate data quality by ensuring that data remains consistent across different records and datasets during data collection and preprocessing [36]. Periodically review data samples manually to identify errors that automated checks might miss and ensure that data values fall within expected ranges and meet specified constraints
		Documentation and Metadata	Keep detailed documentation of data sources, preprocessing steps, and any transformations applied. Maintain metadata that contains the origin and history of the data to ensure traceability and reliability
		Continuous Monitoring and Feedback	Consistently monitor data quality and model performance to detect and resolve emerging issues. Implement feedback loops where model predictions and real-world outcomes are compared to identify discrepancies and improve data quality over time

layer. In Node 2, a team of data scientists and domain experts collaborates to analyze and refine the data, resulting in the second version of the training data. This second version is subsequently stored in the data storage component and ingested to Node 3 for enhancement in the Data Quality Control (DQC) layer. In Node 3, a team of data scientists performs comprehensive quality checks and enhancements on the data, resulting in the third version of the training data.

This iterative process continues until the desired data quality is achieved, with each subsequent version being stored in the data storage component for versioning and traceability. Once the data meets the required quality standards, the final version is fed back into the DCAI layer by the data scientist, where it is used to train an efficient and improved ML model, ready for deployment into production environments. A service-wise overview of the inputs and outputs

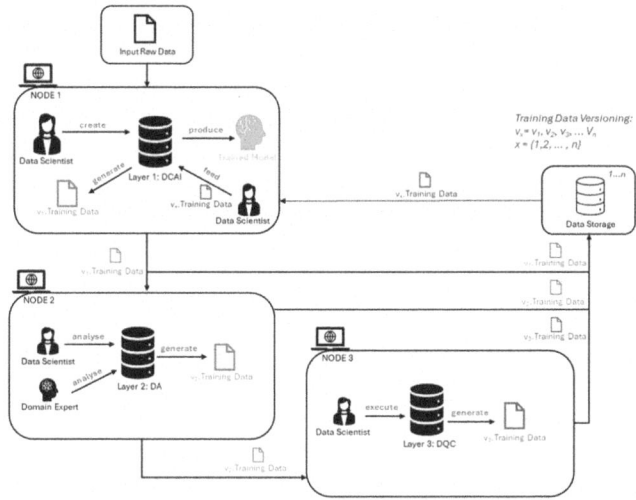

Fig. 4. An exemplary architecture for implementing the LDCAI framework.

to each layer is provided in Table 2, detailing the data flow and transformations occurring within the LDCAI framework's microservice architecture. By adopting a microservice architecture [46], the LDCAI framework benefits from the inherent advantages of modularity, scalability, and independent deployment of individual components. This architectural approach promotes loose coupling between the layers, facilitating easier maintenance, updates, and integration with existing ML infrastructure and pipelines within an organization [47].

Table 2. An overview of inputs and outputs for each layer in the implementation architecture.

Service	Iterations	Layer	Input	Output	Actor
Node 1	1st	DCAI	Raw Data	v1. Training Data, Trained Model	Data Scientist
Node 2	1st	DA	v1. Training Data	v2. Training Data	Data Scientist, Domain Expert
Node 3	1st	DQC	v2. Training Data	v3. Training Data	Data Scientist
Node 1	2nd	DCAI	v2. Training Data/v3. Training Data	Trained Model	Data Scientist
Node 2	2nd	DA	v2. Training Data/v3. Training Data	v4. Training Data	Data Scientist, Domain Expert
Node 3	2nd	DQC	v4. Training Data	v5. Training Data	Data Scientist
...	...	...	...	...	...

4.2 Framework Related Observations

The quality principles outlined in the Data Analysis (DA) and Data Quality Control (DQC) layers of the LDCAI framework in Sect. 4 are not a comprehensive enumeration of all potential issues and solutions related to achieving data quality excellence in AI systems. Instead, they serve as a key illustration of the emerging data-centric AI (DCAI) domain and can be adopted and tailored according to the specific needs and requirements of the DCAI concept implementation. The implementation architecture described in Sect. 4.1 is based on microservice architecture [46], chosen for its inherent flexibility and scalability benefits in deploying ML models [47]. However, this architectural choice is neither fixed nor a prescriptive implementation. The decision for the architecture and design is completely flexible and up to the practitioners for adaptation and deployment according to the specific organizational ML pipeline setup and constraints.

5 Data-Centric AI: Key for Sustainability and Efficiency

High energy consumption of AI systems due to increased computational capabilities and the ever-growing volume of data has become a major concern that requires immediate action. Previously, the main focus was to optimize model architectures and training strategies regardless of energy consumption, whereas the impact of improving the data quality itself has remained an open question [10]. To address this gap, studies from the authors of [4,8,12,14,15,20,22,23] suggest DCAI as a possible solution for handling data quality and implementing efficient AI systems. The authors from [24,26,48,49] suggest that DCAI reduces the energy consumption of AI systems, particularly deep learning models, which require significant computational resources and energy during training and inference. One good example of achieving sustainability in AI systems is the work from the authors of [26], advocating DCAI as a greener AI. They have conducted exploratory studies using six different AI algorithms, a dataset comprising 5,574 data points, and two types of dataset modifications by varying the number of data points and features. They report that by refining the datasets, significant energy savings can be achieved, with reductions of up to 92.16% in energy consumption observed in some cases. Importantly, these energy savings often came at the cost of negligible or no decline in model accuracy. In addition to data modifications, the authors demonstrated incorporating techniques like active learning [50], knowledge transfer [51], dataset distillation [52], data augmentation [44], and curriculum learning [53] into the AI algorithm used can also lead to substantial energy savings of up to two orders of magnitude.

6 Conclusion

LDCAI can handle diversified datasets by mitigating biases and inconsistencies inherent in data leading to equitable and fairer AI systems. By introducing structured data quality checks and iterative improvements, we have presented the LDCAI framework, which expands the existing DCAI pipeline leading to better decision-making based on high-quality data. High-quality data reduces the need for extensive model retraining, saving time and computational resources. With LDCAI's iterative and actionable steps, we ensure that the data-centric approach is rooted in the development process from the start rather than after failure becomes evident, making LDCAI a valuable resource for AI systems. It is crucial to note that while the LDCAI framework provides a structured approach to DCAI implementation, it is not a prescriptive or one-size-fits-all solution. Practitioners are encouraged to tailor the framework's components, principles, and architectural implementation to align with their specific organizational ML infrastructure, needs, data characteristics, and operational constraints, ensuring smooth integration and adoption within their respective environments.

6.1 Future Work

This study focuses on the framework design and establishing a foundation for analysis and quality principles to improve data quality. Continuing our work, we plan to experiment with the LDCAI framework to compare and evaluate different pipeline configurations concerning their efficacy in facilitating AI system development. By implementing the LDCAI framework across various AI use cases, we will evaluate its impact on model performance, development time, and cost. We will also be quantifying the overhead introduced by the additional layers and iterative processes thereby refining and optimizing the LDCAI framework, as well as the potential time and cost savings resulting from improved model performance and reduced rework. This way, we aim to validate the practical applicability of the LDCAI framework and identify areas for further improvement. Ultimately, we aim to establish LDCAI as a robust and widely adopted methodology for developing reliable and high-performing AI systems through a systematic data-centric approach.

References

1. Caldana, V., Silva, F., Oliveira, R., Borin, J.: Internet of things and artificial intelligence applied to predictive maintenance in industry 4.0: a systematic literature review (2021). https://doi.org/10.46254/SA02.20210582
2. Selmy, H.A., Mohamed, H.K., Medhat, W.: Big data analytics deep learning techniques and applications: a survey. Inf. Syst. **120**, 102318 (2024). https://doi.org/10.1016/j.is.2023.102318
3. Singh, S., Hooda, S.: A study of challenges and limitations to applying machine learning to highly unstructured data. In: 2023 7th International Conference On Computing, Communication, Control and Automation (ICCUBEA), pp. 1–6 (2023). https://doi.org/10.1109/ICCUBEA58933.2023.10392115
4. Budach, L., et al.: The effects of data quality on machine learning performance (2022). https://arxiv.org/abs/2207.14529
5. Palanivinayagam, A., Damaševičius, R.: Effective handling of missing values in datasets for classification using machine learning methods. Information **14**(2) (2023). https://doi.org/10.3390/info14020092
6. Balayn, A., Lofi, C., Houben, G.J.: Managing bias and unfairness in data for decision support: a survey of machine learning and data engineering approaches to identify and mitigate bias and unfairness within data management and analytics systems. VLDB J. **30**(5), 739–768 (2021). https://doi.org/10.1007/s00778-021-00671-8

7. Johnson, J., Khoshgoftaar, T.: Survey on deep learning with class imbalance. J. Big Data **6**, 27 (2019). https://doi.org/10.1186/s40537-019-0192-5
8. Hamid, O.: Data-centric and model-centric AI: twin drivers of compact and robust industry 4.0 solutions. Appl. Sci. **13**, 2753 (2023). https://doi.org/10.3390/app13052753
9. DeepLearningAI: A Chat with Andrew on MLOps: From Model-centric to Data-centric AI (2021). https://www.youtube.com/watch?v=06-AZXmwHjo
10. Majeed, A., Hwang, S.O.: Technical analysis of data-centric and model-centric artificial intelligence. IT Prof. **25**(6), 62–70 (2023). https://doi.org/10.1109/MITP.2023.3322410
11. Yu, T., Zhu, H.: Hyper-parameter optimization: a review of algorithms and applications (2020). https://arxiv.org/abs/2003.05689
12. Jain, A., et al.: Overview and importance of data quality for machine learning tasks, pp. 3561–3562 (2020). https://doi.org/10.1145/3394486.3406477
13. Wang, J., Liu, Y., Li, P., Lin, Z., Sindakis, S., Aggarwal, S.: Overview of data quality: examining the dimensions, antecedents, and impacts of data quality. J. Knowl. Econ. **15**(1), 1159–1178 (2024). https://doi.org/10.1007/s13132-022-01096-6
14. Hamid, O.H.: From model-centric to data-centric AI: a paradigm shift or rather a complementary approach? In: 2022 8th International Conference on Information Technology Trends (ITT), pp. 196–199 (2022). https://doi.org/10.1109/ITT56123.2022.9863935
15. Fredriksson, T., Mattos, D.I., Bosch, J., Olsson, H.H.: Data labeling: an empirical investigation into industrial challenges and mitigation strategies. In: Morisio, M., Torchiano, M., Jedlitschka, A. (eds.) PROFES 2020. LNCS, vol. 12562, pp. 202–216. Springer, Cham (2020). https://doi.org/10.1007/978-3-030-64148-1_13
16. Duong, H.T., Le, V.T., Hoang, V.T.: Deep learning-based anomaly detection in video surveillance: a survey. Sensors **23**(11) (2023). https://doi.org/10.3390/s23115024
17. Pansara, R.: Cultivating data quality to strategies, challenges, and impact on decision-making. Int. J. Manage. Educ. Sustain. Dev. **6**(6), 24–33 (2023). https://ijsdcs.com/index.php/IJMESD/article/view/356
18. Jarrahi, M.H., Memariani, A., Guha, S.: The principles of data-centric AI. Commun. ACM **66**(8), 84–92 (2023). https://doi.org/10.1145/3571724
19. Jakubik, J., Vössing, M., Kühl, N., Walk, J., Satzger, G.: Data-centric artificial intelligence. Bus. Inf. Syst. Eng. (2024). https://doi.org/10.1007/s12599-024-00857-8
20. Zha, D., Bhat, Z.P., Lai, K.H., Yang, F., Hu, X.: Data-centric AI: perspectives and challenges (2023). https://arxiv.org/abs/2301.04819
21. Seedat, N., Imrie, F., Schaar, M.: Navigating data-centric artificial intelligence with dc-check: advances, challenges, and opportunities. IEEE Trans. Artif. Intell. **PP**, 1–15 (2023). https://doi.org/10.1109/TAI.2023.3345805
22. Luley, P.P., Deriu, J.M., Yan, P., Schatte, G.A., Stadelmann, T.: From concept to implementation: The data-centric development process for AI in industry. In: 2023 10th IEEE Swiss Conference on Data Science (SDS), pp. 73–76 (2023). https://doi.org/10.1109/SDS57534.2023.00017
23. Kumar, S., Datta, S., Singh, V., Singh, S., Sharma, R.: Opportunities and challenges in data-centric AI. IEEE Access **PP**, 1 (2024). https://doi.org/10.1109/ACCESS.2024.3369417
24. Liang, W., et al.: Advances, challenges and opportunities in creating data for trustworthy AI. Nat. Mach. Intell. **4**(8), 669–677 (2022). https://doi.org/10.1038/s42256-022-00516-1

25. Eyuboglu, S., Karlaš, B., Ré, C., Zhang, C., Zou, J.: DCBench: a benchmark for data-centric AI systems. In: Proceedings of the Sixth Workshop on Data Management for End-To-End Machine Learning, DEEM 2022. Association for Computing Machinery, New York (2022). https://doi.org/10.1145/3533028.3533310
26. Verdecchia, R., Cruz, L., Sallou, J., Lin, M., Wickenden, J., Hotellier, E.: Data-centric green AI an exploratory empirical study. In: 2022 International Conference on ICT for Sustainability (ICT4S), pp. 35–45 (2022). https://doi.org/10.1109/ICT4S55073.2022.00015
27. Maskey, M.: Rethinking AI for science: an evolution from data-driven to data-centric framework. Perspect. Earth Space Sci. **4** (2023). https://doi.org/10.1029/2023CN000222
28. Mazumder, M., et al.: DataPerf: benchmarks for data-centric AI development (2023). https://arxiv.org/abs/2207.10062
29. Whang, S.E., Roh, Y., Song, H., Lee, J.G.: Data collection and quality challenges in deep learning: a data-centric AI perspective. VLDB J. **32**(4), 791–813 (2023). https://doi.org/10.1007/s00778-022-00775-9
30. Polyzotis, N., Zaharia, M.: What can data-centric AI learn from data and ml engineering? (2021). https://arxiv.org/abs/2112.06439
31. Patel, H., Guttula, S., Gupta, N., Hans, S., Mittal, R., Lokesh, N.: A data-centric AI framework for automating exploratory data analysis and data quality tasks. J. Data Inf. Qual. **15** (2023). https://doi.org/10.1145/3603709
32. Bosser, J.D., Sorstadius, E., Chehreghani, M.H.: Model-centric and data-centric aspects of active learning for deep neural networks. In: 2021 IEEE International Conference on Big Data (Big Data). IEEE (2021). https://doi.org/10.1109/bigdata52589.2021.9671795
33. Mallick, A., Hsieh, K., Arzani, B., Joshi, G.: Matchmaker: Data drift mitigation in machine learning for large-scale systems. In: Proceedings of Machine Learning and Systems, vol. 4, pp. 77–94 (2022)
34. Hinder, F., Vaquet, V., Brinkrolf, J., Hammer, B.: Model-based explanations of concept drift. Neurocomputing **555**, 126640 (2023). https://doi.org/10.1016/j.neucom.2023.126640
35. Gong, P., Ma, Y., Li, C., Ma, X., Noh, S.H.: Understand data preprocessing for effective end-to-end training of deep neural networks (2023). https://arxiv.org/abs/2304.08925
36. Dalianis, H.: Evaluation Metrics and Evaluation. In: Dalianis, H. (ed.) Clinical Text Mining, pp. 45–53. Springer, Cham (2018). https://doi.org/10.1007/978-3-319-78503-5_6
37. Murphy, J., Hayden, M., Murphy, B., Ballantine, J.: Approaches to analysis of qualitative research data: a reflection of the manual and technological approaches (2021)
38. Srivastava, D.: An introduction to data visualization tools and techniques in various domains. Int. J. Comput. Trends Technol. **71**, 125–130 (2023). https://doi.org/10.14445/22312803/IJCTT-V71I4P116
39. Kumar, S., Chong, I.: Correlation analysis to identify the effective data in machine learning: prediction of depressive disorder and emotion states. Int. J. Environ. Res. Public Health **15**(12), 2907 (2018)
40. Ehrlinger, L., Rusz, E., Wöß, W.: A survey of data quality measurement and monitoring tools (2019). https://arxiv.org/abs/1907.08138
41. Cai, L., Zhu, Y.: The challenges of data quality and data quality assessment in the big data era. Data Sci. J. **14** (2015). https://doi.org/10.5334/dsj-2015-002

42. Ridzuan, F., Wan Zainon, W.M.N.: A review on data cleansing methods for big data. Procedia Comput. Sci. **161**, 731–738 (2019). https://doi.org/10.1016/j.procs.2019.11.177. The Fifth Information Systems International Conference, 23–24 July 2019, Surabaya, Indonesia
43. Kamalov, F., Moussa, S., Reyes, J.: Data transformation in machine learning: empirical analysis, pp. 115–120 (2023). https://doi.org/10.1109/3ICT60104.2023.10391512
44. Mumuni, A., Mumuni, F.: Data augmentation: a comprehensive survey of modern approaches. Array **16**, 100258 (2022). https://doi.org/10.1016/j.array.2022.100258
45. Rawat, T., Khemchandani, V.: Feature engineering (FE) tools and techniques for better classification performance (2019). https://doi.org/10.21172/ijiet.82.024
46. Roh, S., Jeong, K.M., Cho, H.Y., Huh, E.N.: An efficient microservices architecture for MLOps, pp. 652–654 (2023). https://doi.org/10.1109/ICUFN57995.2023.10201181
47. Ribeiro, J., Figueredo, M., Cacho, N., Lopes, F., de Araujo, A.: A microservice based architecture topology for machine learning deployment (2019). https://doi.org/10.1109/ISC246665.2019.9071708
48. Salehi, S., Schmeink, A.: Data-centric green artificial intelligence: a survey. IEEE Trans. Artif. Intell. **PP**, 1–18 (2023). https://doi.org/10.1109/TAI.2023.3315272
49. Kumar, S., Sharma, R., Singh, V., Tiwari, S., Singh, S.K., Datta, S.: Potential impact of data-centric AI on society. IEEE Technol. Soc. Mag. **42**(3), 98–107 (2023). https://doi.org/10.1109/MTS.2023.3306532
50. Li, D., Wang, Z., Chen, Y., Jiang, R., Ding, W., Okumura, M.: A survey on deep active learning: Recent advances and new frontiers. IEEE Trans. Neural Netw. Learn. Syst. **PP** (2024). https://doi.org/10.1109/TNNLS.2024.3396463
51. Wang, B., et al.: Gap minimization for knowledge sharing and transfer (2023). https://arxiv.org/abs/2201.11231
52. Cazenavette, G., Wang, T., Torralba, A., Efros, A.A., Zhu, J.Y.: Dataset distillation by matching training trajectories (2022). https://arxiv.org/abs/2203.11932
53. Soviany, P., Ionescu, R.T., Rota, P., Sebe, N.: Curriculum learning: a survey (2022). https://arxiv.org/abs/2101.10382

Combining Graph NN and LLM for Improved Text-Based Emotion Recognition

Xinhao Zou and Konstantin Markov(✉)

The University of Aizu, Fukushima 965-8580, Japan
{d8261110,markov}@u-aizu.ac.jp

Abstract. Text-based emotion analysis, an important task in Natural Language Processing (NLP), aims to identify and understand emotional tendencies in text. Recently, given their strong performance in text classification, Graph Neural Networks (GNNs) have been utilized in various emotion recognition studies. They have excellent structural modeling abilities but lack context encoding strength. On the other hand, Large Language Models (LLMs) such as BERT and GPT are specially designed to model the text context. Aiming to utilize both their advantages, we investigated several ways to combine GNNs with LLMs for the emotion recognition task. First, we used BERT to generate embeddings for the graph document nodes. Next, we extended the system to include a description of the input data's emotional content obtained from GPT as an additional node embedding. For experiments and system evaluation, we used the GoEmotions dataset. The results clearly show that combining GNN and LLM improves the emotion classification performance by 20% to 30% compared to when either GNN or LLM is used alone.

Keywords: Graph Neural Network · Large Language Model · Emotion analysis and recognition

1 Introduction

In recent years, emotion analysis has gradually emerged as an important field of Natural Language Processing (NLP) [15]. However, the traditional emotion analysis methods have significant limitations when dealing with complex context and text structure [3]. Sometimes it is difficult to combine context and situation [17,22], and the fixed vocabulary size may affect the performance. Some words often have different meanings in different contexts, which may limit the ability to predict the emotional content precisely [7,8,12]. A wide range of studies in this area use categorical emotion labels which makes the emotion recognition task a supervised classification task. However, continuous [9] or distributed [21] emotion representation has also been explored.

The deep learning technology has significantly transformed the NLP field and as a consequence the text-based emotion recognition methods. Large language models such as BERT [5] and GPT [16] have been utilized in various studies for emotional analysis [1]. On the other hand, Graph Neural Networks (GNN) [25] has also been used for text processing [10,24]. Furthermore, a system combining GNN and BERT has achieved excellent results on the text classification task [13]. Following the same approach, we first built a GNN-based emotion recognition system where document node features are initialized from the corresponding document representation obtained from the BERT model. The results we obtained in terms of emotion recognition accuracy were about 10% higher than if we used only BERT or GNN models separately.

When the text documents are short and as in our case just a single sentence, there might not be enough words to estimate the emotion category reliably. However, using generative LLM such as GPT it is possible to extend the text length and use the LLM power to elaborate on the emotional content of the input sentence. This way we obtain additional text, longer than the original sentence and closely related to its emotional content. Next, we use again the BERT model to extract a representation vector for this additional text explanation. Finally, vectors from both the original text and its explanation are concatenated and used as node features for the GNN. This approach further improved the performance of our system by another 15% accuracy compared to the combined BERT/GNN case.

2 Graph Neural Network

Graph Neural Network (GNN) is a class of deep learning models designed for processing graph-structured data. Entities in graph-structured data are represented as nodes, and the relationships between entities are represented as edges. The goal of GNN is to learn useful representations from graph-structured data and utilize these representations for various tasks. In Graph Neural Networks, the adjacency matrix A is an important tool for describing graph structure. It is a two-dimensional matrix that represents the connection relationships between nodes in the graph. In the adjacency matrix, rows and columns represent nodes, and the matrix elements represent connections between nodes. For a graph, if there is a connection between node i and node j, the elements in the corresponding (i,j) and (j,i) positions in the adjacency matrix typically have a value of 1 or represent the weight of the connection. If there is no connection between the nodes, the corresponding element value is 0.

Each node encompasses a set of defining features. These features can vary based on the application. Edges between nodes might signify connections between entities with similar attributes, indicating some form of association or mutual influence. In this context, these features constitute feature vectors, serving as mathematical representations of nodes that capture various properties and characteristics within the graph structure.

The core idea of GNN is to aggregate neighbor information of nodes through the message-passing mechanism and update the feature representation of nodes. This process typically involves multiple iterations to capture information farther away in the graph. Eventually, each node's feature representation will contain information about its neighbors and more distant nodes.

2.1 Graph Convolutional Network

Consider a graph $G = (V, E)$ where V ($|V| = n$) is a set of nodes, and E is a set of edges. Let $X \in \mathbb{R}^{n \times d}$ be a matrix containing all n nodes d-dimensional features and let A be the adjacency matrix.

GCN [11] is a basic variant of GNN that aggregates node neighbor information through graph convolution operations. The graph convolution operation of GCN can be expressed as:

$$H^{(l+1)} = \sigma(\widetilde{D}^{-\frac{1}{2}} \widetilde{A} \widetilde{D}^{-\frac{1}{2}} H^{(l)} W^{(l)}) \tag{1}$$

where $H^{(l)}$ is the node representation matrix for layer l, $W^{(l)}$ represents the weight matrix of layer l, $\widetilde{A} = A + I$, $\widetilde{D}$ is the degree matrix of $\widetilde{A}$, and σ is a non-linearity such as ReLU [2]. Graphically, the operations in Eq.(1) are depicted in Fig. 1. For the first GCN layer, the node representation is initialized by the node feature matrix, i.e. $H^{(0)} = X$.

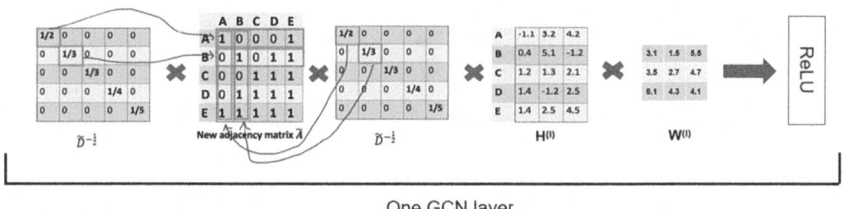

Fig. 1. Single GCN layer operation.

The above Eq.(1) describes the operation of a single GCN layer. Generally, several GCN layers are stacked for a complete GCN. The final layer representation $H^{(L)}$ is fed to a fully connected dense layer to obtain the model output. The overall structure of a multi-layer GCN is shown in Fig. 2:

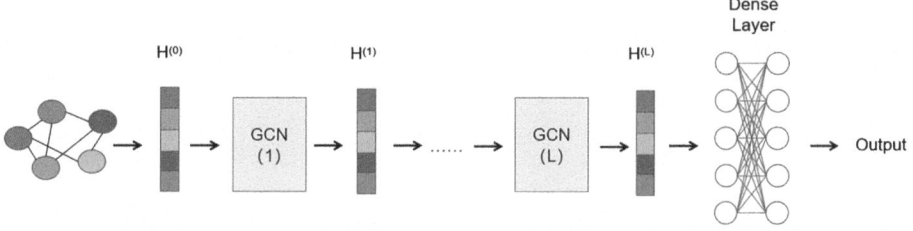

Fig. 2. The structure of a full GCN model.

2.2 GNN for Emotion Recognition

A collection of document texts can be represented by a graph considering the relationship between words and documents they are used in. Such a graph includes two sets of nodes: word nodes and document nodes, each representing one document or one unique word. Edges exist between each document node and word nodes corresponding to the words in that document. Generally, no other edges are needed, but following the approach taken in the TextGCN model [23], we also add weighted edges between word nodes.

With GNN, estimating the document's emotion is to find to which emotional class the corresponding node belongs. This is a semi-supervised task since all the node representations are used during training, but the loss calculation involves only the labeled training nodes. The structure of such a graph is given in Fig. 3 and the adjacency matrix is defined as in Eq.(2).

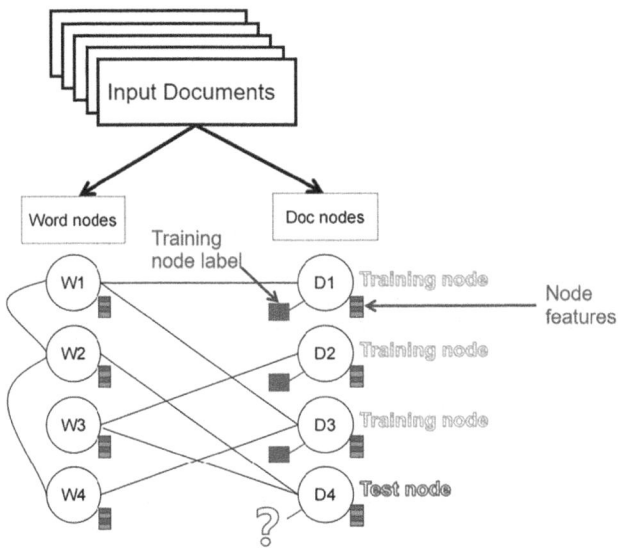

Fig. 3. Graph structure representing documents collection.

$$A_{i,j} = \begin{cases} PPMI(i,j), & i,j \text{ are words and } i \neq j \\ TF\text{-}IDF(i,j), & i \text{ is document}, j \text{ is word} \\ 1, & i = j \\ 0, & \text{otherwise} \end{cases} \quad (2)$$

The word-document edge weights are the word term frequency-inverse document frequency (TF-IDF) value and the word-word edge weights are set to the positive point-wise mutual information (PPMI) value:

$$PPMI(\omega_i, \omega_j) = \max(\log \frac{P(\omega_i, \omega_j)}{P(\omega_i)P(\omega_j)}, 0) \qquad (3)$$

where the word probabilities are estimated from the normalized word counts. In the absence of additional information about words and documents, node features are initialized with an identity matrix, $X = I_{n+m}$, where n is the number of documents and m is the number of unique words, i.e. the vocabulary size.

3 Large Language Models (LLMs)

3.1 BERT-Based Emotion Recognition

BERT is a pre-trained language model built on the Transformer's encoder architecture. There are several variants such as RoBERTa, DistilBERT, etc. which have the same structure but differ in the way they are trained. BERT outputs vector representations of the input words as well as a single vector representing the whole text document, the CLS output. Using this output it is easy to build an emotion classifier as shown in Fig. 4. A couple of linear dense layers take the CLS output and produce the estimated class probabilities using a softmax function. To achieve better performance on the downstream task, in our case - emotion recognition, pre-trained models are usually fine-tuned [19] in a supervised manner using the task data. The same approach can be used with RoBERTa or DistilBERT models [1].

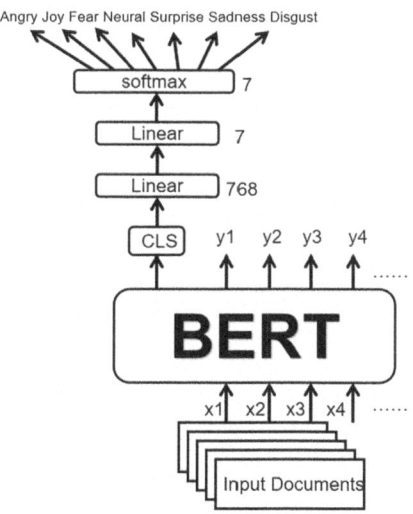

Fig. 4. Schematic diagram of BERT-based emotion recognition model.

3.2 GPT-Based Emotion Recognition

Generative large language models such as GPT3.5 [16] have performed remarkably well in various text-processing tasks. Thus, analyzing the emotional content of a text and classifying it into several categories can be done using LLM alone [20]. The simplest approach would be to create a proper prompt and obtain the LLM answer. The architecture of an LLM-based emotion recognition system is shown in Fig. 5. The LLM is instructed to analyze the input document's emotional content and label it using a set of predefined emotion classes and to output the corresponding label.

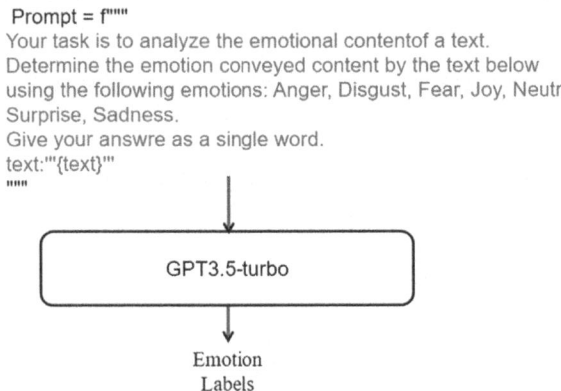

Fig. 5. LLM-based emotion recognition system structure.

4 Combining LLM and GNN

4.1 BERT and GNN

In Sect. 2.2 we introduced the GNN for emotion recognition and defined the initial node features as an identity matrix $X = I_{n+m}$. However, it is possible to use the BERT-derived input text representations as document node features which are much more discriminative and would ensure better performance [13]. In this case, the input feature matrix is defined as:

$$X = \begin{pmatrix} X_{doc} \\ 0 \end{pmatrix}_{(n+m) \times d} \quad (4)$$

where X_{doc} is the matrix of d-dimensional embeddings of all the n documents and the m word node feature vectors are set to zero. The block diagram of such a system is shown in Fig. 6. The BERT model is used similarly as in Sect. 3.1 but without the output softmax layer.

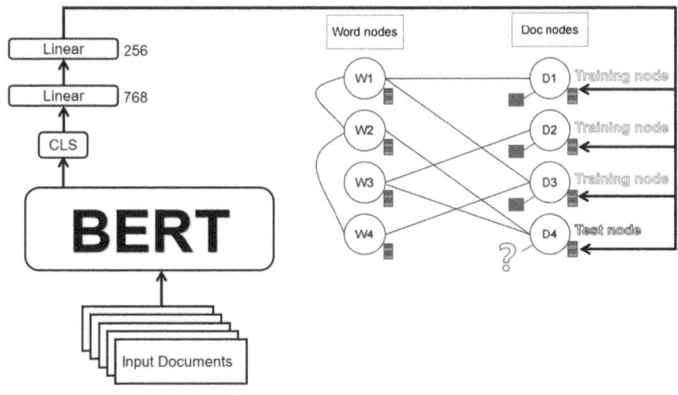

Fig. 6. GNN system with BERT-derived document node features.

To fully utilize the BERT power, the outputs of this system Z_{GNN} can be combined with the outputs of the BERT-only classifier Z_{BERT}:

$$Z = \lambda \cdot Z_{GNN} + (1 - \lambda) \cdot Z_{BERT} \tag{5}$$

where $\lambda \in (0, 1)$ is the linear combination hyper-parameter. Similarly to [13], in our experiments, we use GCN layers (see Sect.2.1) and denote this system as BertGCN.

4.2 BERT, GNN, and GPT

The BERT model can provide highly discriminative document embeddings, but when the input text consists of a single sentence with a few words the classification task becomes more challenging. This problem can be alleviated using the generation abilities of models such as GPT. With an appropriate prompt, based on the GPT's reasoning power, the information about the document's emotional content can be significantly expanded. As shown in Fig. 7, we instruct the model to analyze the emotional content of the input text and provide a reasoning or explanation of its decision within one paragraph. The contextually rich explanations generated by GPT are transformed into representation vectors using BERT in the same way the input documents are processed. Representation vectors for the input and its GPT explanation are further concatenated and used as features for the document nodes schematically depicted in Fig. 8.

5 Experiment

5.1 Dataset

The dataset used in this study is the GoEmotions corpus [4,18]. It consists of 58K Reddit comments annotated manually. The comments are drawn from

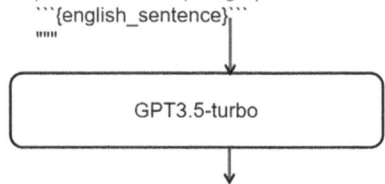

```
prompt = f"""
        Analyze the emotion of the following English text
        from labels below"anger, disgust, fear, joy, neutral,
        surprise, sadness" and explain why you think that. Write
        your explanation in one paragraph: \
        ```{english_sentence}```
        """
```

1 Sentiment Analysis: Based on the given text, the emotion conveyed is joy. The statement expresses a positive sentiment towards food and the joy of not having to cook it oneself. The use of the word "favorite" indicates a strong preference and enjoyment, while the phrase "anything I didn't have to cook myself" suggests a sense of relief and happiness in being able to enjoy food without the effort of cooking. Overall, the text conveys a lighthearted and cheerful emotion, which aligns with the emotion of joy.
2 Sentiment Analysis: Based on the given text, the emotion conveyed is likely to be a combination of anger and sadness. The use of phrases like "if he does off himself" and "everyone will think he's having a laugh screwing with people instead of actually dead" suggests a negative and potentially distressing situation. The mention of someone potentially taking their own life and the insensitivity of others perceiving it as a joke can evoke feelings of anger towards those who may not take such matters seriously. Additionally, the mention of death and the possibility of someone being perceived as not genuinely dead can evoke feelings of sadness and empathy towards the individual in question. Overall, the text conveys a mix of anger and sadness due to the serious nature of the situation and the potential lack of understanding or empathy from others.
3 Sentiment Analysis: The text "WHY THE FUCK IS BAYLESS ISOING" can be categorized as anger. The use of profanity and capitalization suggests a strong negative emotion, which is commonly associated with anger. The phrase "WHY THE FUCK" expresses frustration and annoyance, further indicating anger. Additionally, the mention of "Bayless isoing" implies a specific situation or action that is causing the anger. Overall, the aggressive language and tone used in the text indicate a strong feeling of anger.
4 Sentiment Analysis: Based on the given text "To make her feel threatened," the emotion conveyed is fear. The use of the word "threatened" suggests a sense of danger or potential harm, which typically elicits fear in individuals. The sentence implies that someone intends to cause harm or create a sense of fear in a specific person, indicating a negative and potentially dangerous situation. Therefore, the emotion expressed in this text is fear.

Fig. 7. Generating explanation from GPT3.5 for four input documents.

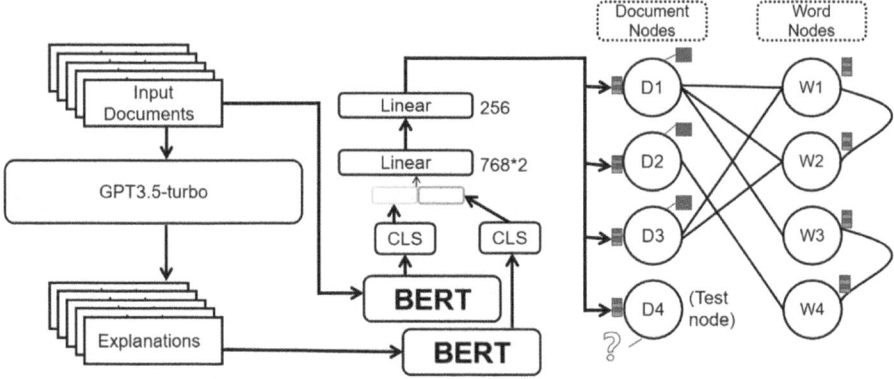

Fig. 8. Integration of GPT generated explanation representations into the BertGCN system.

popular English Reddit subpages and labeled with 27 emotion categories. We mapped those 27 categories to Ekman's 7 emotion classes [3,6] which include anger, disgust, fear, joy, neutral, surprise, and sadness. Comments comprising only a single emoticon were excluded from the dataset reducing its size to about 54K. The dataset was randomly split into the train, validation, and test sets with a ratio of 80:10:10%.
```

5.2 Evaluated Models

We evaluated and compared several configurations of single-model systems as well as combined GNN/LLM models.

- **GCN:** Only graph model with an adjacency matrix that doesn't include word-to-word edges. The nodes feature matrix is set to identity, $X = I$.
- **BERT:** Bert-based emotion classifier described in Sect. 3.1.
- **RoBERTa:** Same as the BERT model, but using the RoBERTa [14].
- **GPT3.5:** GPT3.5 based classification method described in Sect. 3.2.
- **TextGCN:** Same as GCN model, but with PPMI weighted word-to-word edges in the adjacency matrix. The nodes feature matrix is set to identity, $X = I$.
- **BertGCN:** Combined BERT and GCN model described in Sect. 4.1. The nodes feature matrix is defined as in Eq. (4).
- **RoBERTaGCN:** Same as BertGCN, but using RoBERTa instead of BERT.
- **RoBERTa + GPT3.5:** A model where the classification decision is done using only the concatenated representations of the input text and its GPT3.5 explanation. No GNN is applied.
- **RoBERTaGCN + GPT3.5:** RoBERTaGCN model including the GPT3.5 generated explanation representations as described in Sect. 4.2.

Each model's hyper-parameters were manually optimized on the validation dataset. Still, to ensure a fair comparison between them, we fixed some of the parameters such as GCN or linear layer size and number. The BERT-style models included 2 dense layers with sizes of 768 (or 1536) and 256 (or 7) respectively, and the GNN models had 2 GCN layers. A learning rate of 1e-4 for used for the BERT/RoBERTa fine-tuning, while the GNN parameters were updated with a learning rate of 1e-3. Between the linear layers in all models, there was a dropout layer inserted with a dropout probability of 0.5.

5.3 Results

We evaluated the emotion recognition performance of all the models described in the previous section using standard metrics such as Accuracy, Recall, Precision, and F1-score. Our test set consists of about 5000 sentences. The Precision, Recall, and F1-score results of all the models are shown in Table 1. They are divided into two groups: 1) Single model group including GCN, TextGCN, and BERT, RoBERTa, and GPT3.5 models; and 2) Combined model group which includes BertGCN, RoBERTaGCN, RoBERTa+GPT3.5, and RoBERTaGCN+GPT3.5 models. The results clearly show that the combined models achieve much better performance than the single models. The GPT3.5 only result was lower than our expectations, and this is probably due to the 0-shot prompt configuration. However, when used for explanation generation in the combined models GPT3.5 provides a boost in the performance of up to about 15%. The accuracy scores of all the models are given in Fig. 9.

Table 1. Classification results (%) of all the models in terms of Precision, Recall, and F1-score metrics.

Model type		Precision	Recall	F1-score
Single Model	GCN	61.99	52.33	43.81
	GPT3.5	56.23	48.24	50.98
	TextGCN	57.55	57.77	56.38
	BERT	58.57	68.84	65.25
	RoBERTa	60.15	65.82	64.29
Combined	BertGCN	68.19	68.25	67.96
	RoBERTaGCN	68.31	68.49	68.09
	RoBERTa + GPT3.5	81.76	81.78	81.71
	RoBERTaGCN + GPT3.5	**84.44**	**84.51**	**84.36**

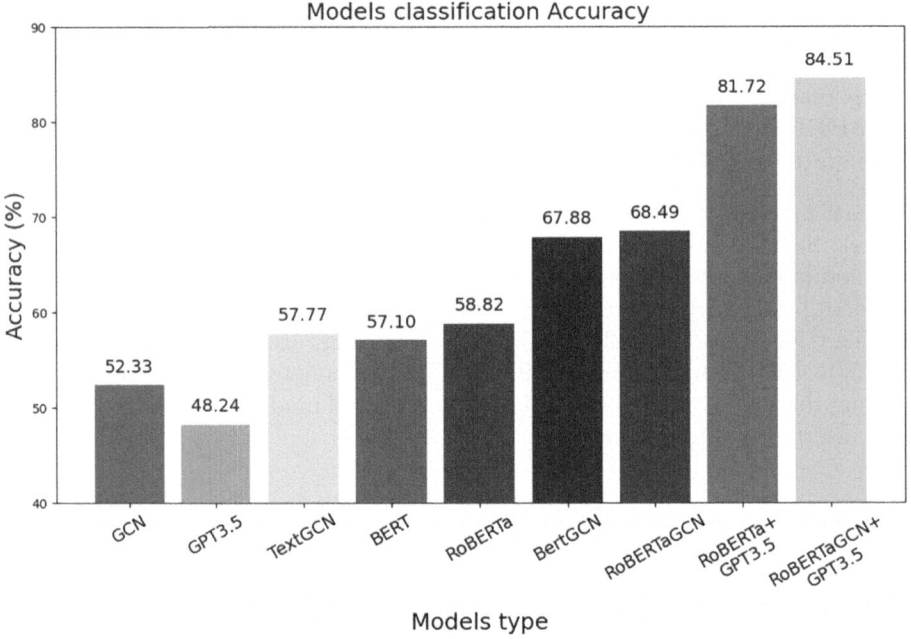

Fig. 9. Emotion classification accuracy (%) of all the models.

6 Conclusion

In this study, we investigated the text-based emotion recognition performance of various GNN-based and LLM-based models as well as several ways to combine them. For the model training and evaluation, we used the GoEmotion dataset reducing the original 27 emotion classes to Ekman's 7 basic categories. All the GNN-based models were trained from scratch while all the LLM-based models but the GPT3.4 were fine-tuned. Evaluation experiments clearly showed that combining GCN and LLM models is effective and improves the performance by up to 20%. Single GCN model results were slightly worse than single BERT or RoBERTa, but higher than GPT3.5 which we attribute to the simple prompt we used. However, when the representations of the GPT3.5 derived explanations were included in the combined models a significant boost in the performance was observed. The reason might be the fact that those explanations were much longer than the input texts and their vector embeddings became easier to discriminate.

There are probably other possible ways to combine the Graph Neural Networks with large language models which based on our findings would also lead to improved performance. The main goal of such approaches is to be able to efficiently merge different views of the input document's information content.

References

1. Adoma, A.F., Henry, N.M., Chen, W.: Comparative analyses of BERT, RoBERTa, DistilBERT, and XLNet for text-based emotion recognition. In: 2020 17th International Computer Conference on Wavelet Active Media Technology and Information Processing (ICCWAMTIP), pp. 117–121. IEEE (2020)
2. Agarap, A.F.: Deep learning using Rectified Linear Units (ReLU). arXiv preprint arXiv:1803.08375 (2018)
3. Coppini, S., Lucifora, C., Vicario, C.M., Gangemi, A.: Experiments on real-life emotions challenge Ekman's model. Sci. Rep. **13**(1), 9511 (2023)
4. Demszky, D., Movshovitz-Attias, D., Ko, J., Cowen, A., Nemade, G., Ravi, S.: GoEmotions: a dataset of fine-grained emotions. arXiv preprint arXiv:2005.00547 (2020)
5. Devlin, J., Chang, M.W., Lee, K., Toutanova, K.: BERT: pre-training of deep bidirectional transformers for language understanding. arXiv preprint arXiv:1810.04805 (2018)
6. Ekman, P., et al.: Basic emotions. Handb. Cogn. Emot. **98**(45–60), 16 (1999)
7. Fung, P., Ngai, G., Yang, Y., Chen, B.: A maximum-entropy Chinese parser augmented by transformation-based learning. ACM Trans. Asian Lang. Inf. Process. (TALIP) **3**(2), 159–168 (2004)
8. Gao, S., Sethi, A., Agarwal, S., Chung, T., Hakkani-Tur, D.: Dialog state tracking: a neural reading comprehension approach. arXiv preprint arXiv:1908.01946 (2019)
9. Gunes, H., Pantic, M.: Automatic, dimensional and continuous emotion recognition. Int. J. Synthetic Emot. (IJSE) **1**(1), 68–99 (2010)
10. Huang, L., Ma, D., Li, S., Zhang, X., Wang, H.: Text level Graph Neural Network for text classification. arXiv preprint arXiv:1910.02356 (2019)
11. Kipf, T.N., Welling, M.: Semi-supervised classification with graph convolutional networks. arXiv preprint arXiv:1609.02907 (2016)
12. Li, Y., Chan, J., Peko, G., Sundaram, D.: Mixed emotion extraction analysis and visualisation of social media text. Data Knowl. Eng. **148**, 102220 (2023)
13. Lin, Y., et al.: BertGCN: transductive text classification by combining GCN and BERT. arXiv preprint arXiv:2105.05727 (2021)
14. Liu, Y., et al.: RoBERTa: a robustly optimized BERT pretraining approach. arXiv preprint arXiv:1907.11692 (2019)

15. Nandwani, P., Verma, R.: A review on sentiment analysis and emotion detection from text. Soc. Netw. Anal. Min. **11**(1), 81 (2021)
16. OpenAI: GPT-3.5: Language model (2021). https://chat.openai.com
17. Shi, W., Xue, G., He, S.: Literature review of network public opinion research from the perspective of sentiment. Doc. Inf. Knowl. **39**(1), 105–118 (2022)
18. Singh, G., Brahma, D., Rai, P., Modi, A.: Text-based fine-grained emotion prediction. IEEE Trans. Affect. Comput. (2023)
19. Sun, C., Qiu, X., Xu, Y., Huang, X.: How to fine-tune BERT for text classification? In: Chinese Computational Linguistics: 18th China National Conference, CCL 2019, Kunming, China, 18–20 October 2019, pp. 194–206. Springer (2019)
20. Wake, N., Kanehira, A., Sasabuchi, K., Takamatsu, J., Ikeuchi, K.: Bias in emotion recognition with ChatGPT. arXiv preprint arXiv:2310.11753 (2023)
21. Wang, X., Zong, C.: Distributed representations of emotion categories in emotion space. In: Proceedings of the 59th Annual Meeting of the Association for Computational Linguistics and the 11th International Joint Conference on Natural Language Processing (Volume 1: Long Papers), pp. 2364–2375 (2021)
22. Xie, Y., Li, J., Pu, P.: Uncertainty and surprisal jointly deliver the punchline: exploiting incongruity-based features for humor recognition. arXiv preprint arXiv:2012.12007 (2020)
23. Yao, L., Mao, C., Luo, Y.: Graph convolutional networks for text classification. In: Proceedings of the AAAI Conference on Artificial Intelligence, vol. 33, pp. 7370–7377 (2019)
24. Zhang, X., Zhao, J., LeCun, Y.: Character-level Convolutional Networks for text classification. In: Advances in Neural Information Processing Systems, vol. 28 (2015)
25. Zhou, J., et al.: Graph neural networks: a review of methods and applications. AI Open **1**, 57–81 (2020)

A Novel Study on Modelling and Adaptive Optimal Control of a Tubular Reactor Based on Gaussian Processes

Alexandra Grancharova[1]($^{\boxtimes}$), Junhong Xie[1], and Juš Kocijan[2,3]

[1] Department of Industrial Automation, University of Chemical Technology and Metallurgy Sofia, Sofia, Bulgaria
`alexandra.grancharova@abv.bg`
[2] Department of Systems and Control, Jozef Stefan Institute, Ljubljana, Slovenia
`jus.kocijan@ijs.si`
[3] University of Nova Gorica, Rožna Dolina, Slovenia

Abstract. Number of engineering systems can be characterized as complex since they have a dynamic and nonlinear behaviour incorporating a stochastic uncertainty. On the other hand, as a machine learning method, Gaussian processes (GP) provide a practical, probabilistic approach to learning in kernel machines. This makes them very suitable for obtaining probabilistic, nonparametric black-box models of stochastic nonlinear dynamic systems. In this paper, a novel study on the modeling and adaptive optimal control of a tubular reactor is made by using Gaussian processes. Such reactor is a typical example of a nonlinear distributed parameters system, whose first-principles dynamic models consists of partial differential equations. The purpose is to obtain a nonlinear autoregressive models with exogenous input (NARX) of the output concentration and temperature of the reactor by applying a GP modeling approach. The identified surrogate models are then used to design an adaptive model predictive controller to achieve optimal performance of the reactor despite of the stochastic changes in the feed temperature. The performance of the adaptive MPC based on GP models is studied by simulation experiments.

Keywords: Gaussian processes · Machine learning · Adaptive optimal control · Model predictive control · Tubular reactors

1 Introduction

Over the last decades there has been a great activity in the "kernel machines" area of *machine learning* as a subfield of the *artificial intelligence*. This includes mainly the work on support vector machines as well as the application of Gaussian process models to machine learning tasks [18]. As a *machine learning method*, *Gaussian processes* provide a principled, practical, probabilistic approach to learning in kernel machines [18]. As it is shown in [18], Gaussian processes are mathematically equivalent to many well known models, including Bayesian linear models, spline models, large neural networks, and are closely related to others, such as support vector machines. Moreover, Gaussian process models may be easier to handle and interpret than their conventional counterparts, such as e.g. neural networks.

Number of engineering systems can be characterized as complex since they have a dynamic and nonlinear behaviour incorporating a stochastic uncertainty. Dynamic-systems control design utilizes various kinds of computational intelligence methods for model development that result in so-called black-box models. Gaussian-process (GP) models provide a probabilistic, nonparametric modelling approach for black-box identification of nonlinear dynamic systems. They can highlight areas of the input space where model-prediction quality is poor, due to the lack of data or its complexity, by indicating the higher variance around the predicted mean. This property can be incorporated in the closed-loop control design. Gaussian-process models contain noticeably less coefficients to be optimized than parametric models that are frequently used in control design. This modelling method is not suggested as a replacement to any existing systems identification method, but rather as a complementary approach to modeling.

The use of Gaussian processes in the modelling of dynamic systems is a relatively recent development e.g. [8, 10, 21]. A number of control design approaches have been developed based on GP models of the dynamic systems. This includes the inverse dynamics control method [15], stochastic model predictive control (MPC) by using online optimization [1, 5, 9, 12–14], learning a GP approximation of an MPC [19], explicit stochastic MPC [2, 3], distributed stochastic MPC of interconnected systems [4], adaptive control [16, 17, 20]. In a broad sense, a recent review on the machine learning approaches used in the control of (bio) chemical manufacturing processes can be found in [7].

In this paper, a novel study on the modeling and adaptive optimal control of a tubular reactor is made by using *Gaussian processes as a machine learning approach*. Such reactor is a typical example of a nonlinear distributed parameters system, whose first-principles dynamic models consists of partial differential equations representing the material and energy balance in the spatiotemporal space. In this paper, the purpose is to obtain a nonlinear autoregressive models with exogenous input (NARX) of the output concentration and temperature of the reactor by applying a Gaussian processes modeling approach. The identified surrogate models are then used to design an adaptive model predictive controller to achieve optimal performance of the reactor despite of the changing feed temperature. This is a preliminary study and based on its results we are planning next steps, among them some tests on real-life data and plant.

2 Preliminaries on Modelling with Gaussian Processes

A Gaussian process is an example of using a flexible, probabilistic, nonparametric model, which directly provides us with uncertainty predictions. In [18], its use and properties for modeling are reviewed. A Gaussian process is a collection of random variables with a joint multivariate Gaussian distribution. Assuming a relationship of the form $y = f(z) + \xi$ (with ξ being a stochastic noise) between input $z \in \mathbb{R}^D$ and output $y \in \mathbb{R}$, we have $y(1), y(2), \ldots, y(M) \sim \mathcal{N}(0, \Sigma)$, where $\Sigma_{pq} = \text{Cov}(y(p), y(q)) = C(z(p), z(q))$ gives the covariance between the output points $y(p)$ and $y(q)$ corresponding to the input points $z(p)$ and $z(q)$. Thus, the mean $\mu(z)$ (usually assumed to be zero) and the covariance function $C(z(p), z(q))$ fully specify the Gaussian process. Note that the covariance function $C(z(p), z(q))$ can be any function with the property that generates a positive definite covariance matrix. Forms of covariance functions suitable for different applications can be found in [18]. A possible covariance function is the isometric rational-quadratic function:

$$C(z(p), z(q)|\theta) = \sigma_f^2 \left(1 + \frac{[z(p) - z(q)]^T [z(p) - z(q)]}{2a\sigma_l^2}\right)^{-a} \tag{1}$$

where σ_l is the characteristic length scale, σ_f is the signal standard deviation and a is a positive hyperparameter. For a given problem, the hyperparameters θ of the covariance function are identified using the data at hand.

Consider a set of M D-dimensional input vectors $\mathbf{Z} = [z(1), z(2), \ldots, z(M)]^T$ and an output data vector $Y = [y(1), y(2), \ldots, y(M)]^T$. Based on the data $(\mathbf{Z}, Y)$, and given a new input vector z^*, we wish to estimate the probability distribution of the corresponding output y^*. Unlike other models, there is no model parameter determination within a fixed model structure. With this model, most of the effort consists of *tuning* the hyperparameters of the covariance function. One of the approaches is by maximizing the log-likelihood of the parameters, i.e. optimally determining them from the evidence (or marginal distribution) of the GP posterior. This is called the empirical Bayes or Type II maximum likelihood optimization. The maximum a posteriori (MAP) estimate of the hyperparameters equals the maximum marginal likelihood estimate of the evidence (not parameters) of the GP posterior (for more details, refer to [18] and [10]). A covariance matrix $\mathbf{K}$ of size $M \times M$ is determined based on the training set $\mathbf{Z}$. As mentioned, the aim is to estimate the probability distribution of the corresponding output y^* at some new input vector z^*. For a new test input z^*, the posterior distribution of the corresponding output is $y^*|z^*, (\mathbf{Z}, Y)$ and is Gaussian, with mean and variance [18]:

$$\begin{aligned} E\{y^*\} &= \mu(z^*) = c(z^*)^T \mathbf{K}^{-1} Y \\ \text{var}\{y^*\} &= \sigma^2(z^*) = c_0(z^*) - c(z^*)^T \mathbf{K}^{-1} c(z^*) + v_0 \end{aligned} \tag{2}$$

where $c(z^*) = [C(z(1), z^*), \ldots, C(z(M), z^*)]^T$ is the vector of covariances between the test and training cases, $c_0(z^*) = C(z^*, z^*)$ and is the covariance between the test input and itself.

Gaussian processes can model static nonlinearities and can, therefore, be used for the modelling of dynamic systems if delayed input and output signals are used as regressors [10]. In such cases, a nonlinear autoregressive model with exogenous input (NARX) is considered, where the current predicted output depends on previously estimated outputs and previous inputs:

$$\begin{aligned} z(k) &= [\hat{y}(k-1), \ldots, \hat{y}(k-m), u(k-1), \ldots, u(k-m)]^T \\ \hat{y}(k) &= \tilde{f}(z(k)) + \eta(k) \end{aligned} \tag{3}$$

where k denotes a consecutive number of data samples, m is a given lag, and $\eta(t)$ is the prediction error.

3 Dynamic Modelling of a Tubular Reactor by Using Gaussian Processes

3.1 Description of the Tubular Reactor

A tubular reactor is considered, in which a first-order irreversible reaction occurs. The reaction is exothermic, and a cooling agent (water) with flowrate u and temperature $\tilde{T}_c$ runs through the reactor jacket (Fig. 1). It is assumed that the tubular reactor is of a plug flow type, i.e. it has cylindrical geometry, the fluid is perfectly mixed in the radial direction, but there is no mixing in the axial direction.

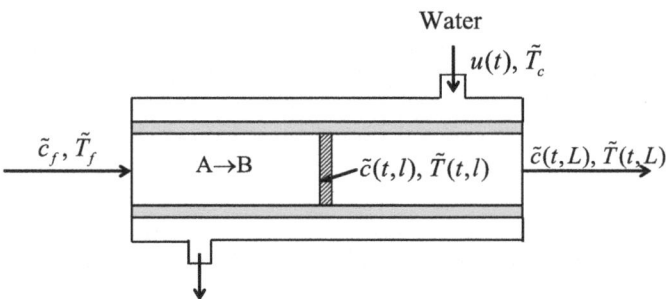

Fig. 1. Jacketed tubular reactor with length L.

The mass and heat balance of the reactor in dynamic mode is expressed through the dimensionless concentration $\tilde{c}$ and temperature $\tilde{T}$ as follows:

$$\frac{\partial \tilde{c}}{\partial t} = -v\frac{\partial \tilde{c}}{\partial l} - k_0 e^{-\frac{E}{\tilde{T}}} \tilde{c} \quad (4)$$

$$\frac{\partial \tilde{T}}{\partial t} = -v\frac{\partial \tilde{T}}{\partial l} + k_0 e^{-\frac{E}{\tilde{T}}} \tilde{c} - \alpha u(\tilde{T} - \tilde{T}_c) \quad (5)$$

where t is the continuous time, l is the spatial coordinate, and the dimensionless quantities are defined as:

$$\tilde{c} = \frac{c}{c_f}, \quad \tilde{T} = \frac{T}{Jc_f}, \quad \tilde{T}_c = \frac{T_c}{Jc_f}, \quad \tilde{T}_f = \frac{T_f}{Jc_f} \quad (6)$$

Here, c and c_f are the concentrations in the reactor and of the feed stream, T, T_c and T_f are the temperatures respectively in the reactor, of the cooling stream and of the feed stream, J is the heat of the chemical reaction. The reaction rate is given by:

$$r = k_0 e^{-\frac{E}{\tilde{T}}} \tilde{c} \quad (7)$$

where k_0 is the pre-exponential multiplier and E is the activation energy of the reaction. In (4)–(5), v is the linear speed of flow through the reactor and α is the heat transfer coefficient. The coolant flowrate u is the control input, and it is constrained to be in the interval:

$$0 \leq u(t) \leq 600 \tag{8}$$

The model (4)–(7) of the plug flow reactor is adapted from the model of a continuous stirred tank reactor in [6]. The values of the parameters are: $c_f = 1$, $T_f = 300$, $v = 0.1$, $L = 0.6$, $E = 25.2$, $k_0 = 300$, $T_c = 290$, $\alpha = 1.95 \cdot 10^{-4}$, $J = 100$ (most of them taken from [6]). The dynamic model (4)–(5) represents a system of nonlinear first-order partial differential equations (PDEs), and a numerical method has to be applied to solve it. Here, the method of lines [11] and [22] is used to solve the model (4)–(5). The idea is to discretize the model by approximating the spatial derivatives by finite differences. In this way, the PDEs (4)–(5) are transformed into a set of coupled ordinary differential equations (ODEs) that can be solved by using some of the methods for numerical solution of ODEs.

3.2 Gaussian Process Models of Reactor Dynamics

The analytical model described in Sect. 3.1 is used to generate data about the dynamic operation of the tubular reactor. By varying the control input u (the coolant flowrate), the PDEs (4)–(5) are solved and data about the reagent concentration and the temperature at the output of the reactor (i.e. for $l = L$) are obtained. In practice, such data are obtained by experiments. Since the dynamics of reactor is sensitive to the variations of feed flow parameters, the model (4)–(5) is solved for different values of the feed temperature $\tilde{T}_f$ (note that the feed concentration of the only reagent is always $c_f = 1$). Let $\mathcal{T}_f$ be the following set of ten values of $\tilde{T}_f$:

$$\mathcal{T}_f = \{\tilde{T}_{f,i} = 2.9 + 0.067i, \ i = 1, 2, \ldots, 10\} \tag{9}$$

Based on the available data, ten sets of NARX models describing the dynamics of the concentration $\tilde{c}_L$ and temperature $\tilde{T}_L$ at the output of the reactor are identified corresponding to values $\tilde{T}_{f,i} \in \mathcal{T}_f$ of the feed temperature:

$$\tilde{c}_L(k) = f_i(\tilde{T}_L(k-1), \ldots, \tilde{T}_L(k-n_T), u(k-1), \ldots, u(k-n_u)|\theta_{f_i}) \tag{10}$$

$$\tilde{T}_L(k) = g_i(\tilde{c}_L(k-1), \ldots, \tilde{c}_L(k-m_c), \tilde{T}_L(k-1), \ldots, \tilde{T}_L(k-m_T),$$
$$u(k-1), \ldots, u(k-m_u)|\theta_{g_i}) \tag{11}$$

$$i = 1, 2, \ldots, 10$$

Here, θ_{f_i} and θ_{g_i} are the hyperparameters of the i-th models (10)–(11) and the lags in these models are:

$$n_T = 6, \ n_u = 4, \ m_c = 9, \ m_T = 13, \ m_u = 9 \tag{12}$$

The set of models (10)–(11) represents a surrogate model of the reactor dynamics. Such a grid of independent models is simpler to design, train and interpret than one complex model that should incorporate the effect of $\tilde{T}_f$ on the reactor dynamics.

Each submodel of the set of independent models was selected as a Gaussian process (GP) model with the NARX structure given above, with zero-mean function and the isometric squared-exponential covariance function (1). This structure was obtained with 4-fold cross-validation using the Regression Learner application in Matlab.

A random number generator generated a test control input signal with normal distribution and rate of change different from those for the identification signal. The responses of the Gaussian process models for concentration and temperature obtained for $\tilde{T}_f = 3$ and for $\tilde{T}_f = 3.3$ are shown in Figs. 2 and 3, respectively.

The individual submodels might not have an optimal structure, but we decided to stick to equal lags for all models in the grid.

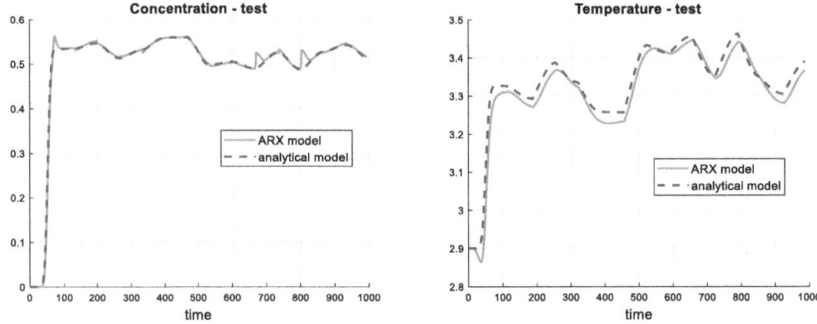

Fig. 2. Response of the GP models for $\tilde{T}_f = 3$.

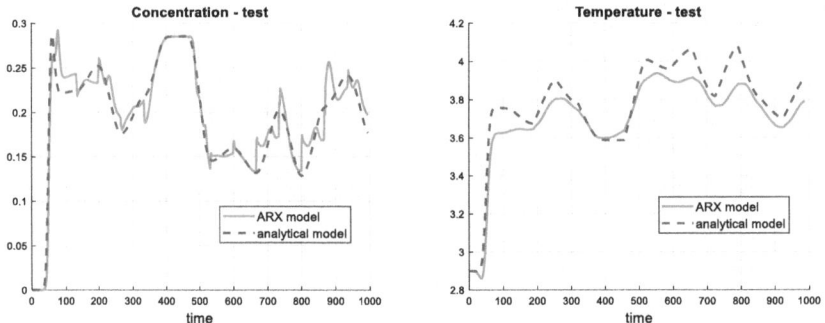

Fig. 3. Response of the GP models for $\tilde{T}_f = 3.3$.

4 Adaptive Optimal Control of the Tubular Reactor Based on the GP Models

4.1 Formulation of a Supervisory Optimal Control Problem and Scheme of the Adaptive System

By defining the vectors of regressors:

$$z_{\tilde{c}}(k) = [\tilde{T}_L(k), \tilde{T}_L(k-1), \ldots, \tilde{T}_L(k-n_T+1), u(k-1), \ldots, u(k-n_u+1)] \tag{13}$$

$$z_{\tilde{T}}(k) = [\tilde{c}_L(k), \tilde{c}_L(k-1), \ldots, \tilde{c}_L(k-m_c+1), \tilde{T}_L(k), \tilde{T}_L(k-1), \ldots, \tilde{T}_L(k-m_T+1),$$
$$u(k-1), \ldots, u(k-m_u+1)] \tag{14}$$

we obtain the following set of compact prediction models for the output concentration and temperature from the models (10)–(11):

$$\tilde{c}_L(k+1) = f_i(z_{\tilde{c}}(k), u(k)|\theta_{f_i}) \tag{15}$$

$$\tilde{T}_L(k+1) = g_i(z_{\tilde{T}}(k), u(k)|\theta_{g_i}) \tag{16}$$

$$i = 1, 2, \ldots, 10$$

Consider the optimal regulation problem for the tubular reactor, where the goal is to steer the output concentration and temperature to the optimal set-point values $\tilde{c}_{L,\text{sp}}$ and $\tilde{T}_{L,\text{sp}}$, which implicitly depend on the feed temperature $\tilde{T}_f$:

$$\tilde{c}_{L,\text{sp}} = h_1(\tilde{T}_f), \quad \tilde{T}_{L,\text{sp}} = h_2(\tilde{T}_f) \tag{17}$$

For $\tilde{T}_{f,i} \in \mathcal{T}_f$, the optimal set-point values $\tilde{c}_{L,\text{sp},i}$, $\tilde{T}_{L,\text{sp},i}$ are determined by solving the steady-state optimization problem:

Problem P1:

$$\tilde{c}_{L,\text{sp},i} = \min_u \tilde{c}_{L,\text{steady},i}(u) \tag{18}$$

subject to the steady state model:

$$\tilde{c}_{L,\text{steady},i}(u) = \lim_{k \to \infty} \tilde{c}_L(k+1) = \lim_{k \to \infty} f_i(z_{\tilde{c}}(k), u|\theta_{f_i}) \tag{19}$$

$$\tilde{T}_{L,\text{steady},i}(u) = \lim_{k \to \infty} \tilde{T}_L(k+1) = \lim_{k \to \infty} g_i(z_{\tilde{T}}(k), u|\theta_{g_i}) \tag{20}$$

with given initial regressor vectors $z_{\tilde{c}}(0)$, $z_{\tilde{T}}(0)$ and respecting the constraint:

$$0 \leq u \leq 600 \tag{21}$$

The optimal solution to Problem P1 is denoted $u_{\text{steady},i}$, and the corresponding optimal set-point values for the output concentration and temperature are:

$$\tilde{c}_{L,\text{sp},i} = \tilde{c}_{L,\text{steady},i}(u_{\text{steady},i}) \tag{22}$$

$$\tilde{T}_{L,\text{sp},i} = \tilde{T}_{L,\text{steady},i}(u_{\text{steady},i}) \qquad (23)$$

At the current time k, denote with $\tilde{c}_L(k)$ and $\tilde{T}_L(k)$ the measured output concentration and temperature of the reactor and with $\tilde{T}_f(k)$ the measured feed temperature. Also, let i^* be the index of the value $\tilde{T}_{f,i} \in \mathcal{T}_f$ that is most close to $\tilde{T}_f(k)$, i.e.:

$$\tilde{T}_{f,i^*} = \arg \min_{\tilde{T}_{f,i} \in \mathcal{T}_f} |\tilde{T}_{f,i} - \tilde{T}_f(k)| \qquad (24)$$

Then, the respective optimal set-point values are denoted with $\tilde{c}_{L,\text{sp},i^*}$, $\tilde{T}_{L,\text{sp},i^*}$ and the control input steady-state value with u_{steady,i^*}. Suppose that at the current time k the regressor vectors $z_{\tilde{c}}(k) = z_{\tilde{c},k|k}$ and $z_{\tilde{T}}(k) = z_{\tilde{T},k|k}$ are known and the control inputs $u(k+j) = u_{k+j}, j = 0, 1, \ldots, N-1$ are given (here N denotes the prediction horizon). Then, the model (15)–(16) can be used to obtain the predicted output concentration $\tilde{c}_{L,k+j+1|k}$ and temperature $\tilde{T}_{L,k+j+1|k}, j = 0, 1, \ldots, N-1$ through iterative one-step ahead predictions, where at each step the predicted output value is fed back to the regressor vectors:

$$\tilde{c}_{L,k+j+1|k} = f_{i^*}(z_{\tilde{c},k+j|k}, u_{k+j}|\theta_{f_{i^*}}) \qquad (25)$$

$$\tilde{T}_{L,k+j+1|k} = g_{i^*}(z_{\tilde{T},k+j|k}, u_{k+j}|\theta_{g_{i^*}}) \qquad (26)$$

In order to avoid a possible steady-state offset of the nonlinear model predictive controller (NMPC) due to the inaccuracy of the prediction with the NARX models, two more output variables (denoted $e_{\tilde{c}}$ and $e_{\tilde{T}}$) are predicted, which represent the integral regulation errors for the output concentration and temperature:

$$e_{\tilde{c},k+j+1|k} = e_{\tilde{c},k+j|k} + (\tilde{c}_{L,k+j|k} - \tilde{c}_{L,\text{sp},i^*}) \qquad (27)$$

$$e_{\tilde{T},k+j+1|k} = e_{\tilde{T},k+j|k} + (\tilde{T}_{L,k+j|k} - \tilde{T}_{L,\text{sp},i^*}) \qquad (28)$$

The initial values of these errors are set to zeros, i.e. $e_{\tilde{c}}(0) = 0$, $e_{\tilde{T}}(0) = 0$. Denote with $n_{\max}$ the maximal lag among the lags in (12). Then, for the current regressor vectors $z_{\tilde{c}}(k)$ and $z_{\tilde{T}}(k)$ (where $k \geq n_{\max}$), and the current values of the integral errors denoted $e_{\tilde{c}}(k)$ and $e_{\tilde{T}}(k)$, the regulation NMPC solves the optimization problem:

Problem P2:

$$V^*(z_{\tilde{c}}(k), z_{\tilde{T}}(k)) = \min_U J(U, z_{\tilde{c}}(k), z_{\tilde{T}}(k)) \qquad (29)$$

subject to $z_{\tilde{c},k|k} = z_{\tilde{c}}(k)$, $z_{\tilde{T},k|k} = z_{\tilde{T}}(k)$, $e_{\tilde{c},k|k} = e_{\tilde{c}}(k)$, $e_{\tilde{T},k|k} = e_{\tilde{T}}(k)$ and:

$$u_{\min} \leq u_{k+j} \leq u_{\max}, \quad j = 0, 1, \ldots, N-1 \qquad (30)$$

$$\tilde{c}_{L,k+j+1|k} = f_{i^*}(z_{\tilde{c},k+j|k}, u_{k+j}|\theta_{f_{i^*}}), \quad j = 0, 1, \ldots, N-1 \qquad (31)$$

$$\tilde{T}_{L,k+j+1|k} = g_{i^*}(z_{\tilde{T},k+j|k}, u_{k+j}|\theta_{g_{i^*}}), \quad j = 0, 1, \ldots, N-1 \qquad (32)$$

$$e_{\tilde{c},k+j+1|k} = e_{\tilde{c},k+j|k} + (\tilde{c}_{L,k+j|k} - \tilde{c}_{L,\text{sp},i^*}), \quad j = 0, 1, \ldots, N-1 \quad (33)$$

$$e_{\tilde{T},k+j+1|k} = e_{\tilde{T},k+j|k} + (\tilde{T}_{L,k+j|k} - \tilde{T}_{L,\text{sp},i^*}), \quad j = 0, 1, \ldots, N-1 \quad (34)$$

$$z_{\tilde{c},k+j|k} = [\tilde{T}_{L,k+j|k}, \tilde{T}_{L,k+j-1|k}, \ldots, \tilde{T}_{L,k+j-n_T+1|k}, u_{k+j-1}, \ldots, u_{k+j-n_u+1}] \quad (35)$$

$$z_{\tilde{T},k+j|k} = [\tilde{c}_{L,k+j|k}, \tilde{c}_{L,k+j-1|k}, \ldots, \tilde{c}_{L,k+j-m_c+1|k}, \tilde{T}_{L,k+j|k}, \tilde{T}_{L,k+j-1|k}, \ldots,$$
$$\tilde{T}_{L,k+j-m_T+1|k}, u_{k+j-1}, \ldots, u_{k+j-m_u+1}] \quad (36)$$

with $U = [u_k, u_{k+1}, \ldots, u_{k+N-1}]$ and the cost function given by:

$$J(U, z_{\tilde{c}}(k), z_{\tilde{T}}(k)) = \sum_{k=0}^{N-1} J_S(\tilde{c}_{L,k+j|k}, \tilde{T}_{L,k+j|k}, e_{\tilde{c},k+j|k}, e_{\tilde{T},k+j|k}, u_{k+j})$$
$$+ J_T(\tilde{c}_{L,k+N|k}, \tilde{T}_{L,k+N|k}, e_{\tilde{c},k+N|k}, e_{\tilde{T},k+N|k}) \quad (37)$$

Here, N is a finite horizon, J_S denotes the stage cost defined as:

$$J_S(\tilde{c}_{L,k+j|k}, \tilde{T}_{L,k+j|k}, e_{\tilde{c},k+j|k}, e_{\tilde{T},k+j|k}, u_{k+j}) = q_1(\tilde{c}_{L,k+j|k} - \tilde{c}_{L,\text{sp},i^*})^2$$
$$+ q_2(\tilde{T}_{L,k+j|k} - \tilde{T}_{L,\text{sp},i^*})^2 + q_3 e_{\tilde{c},k+j|k}^2 + q_4 e_{\tilde{T},k+j|k}^2 + q_5(u_{k+j} - u_{\text{steady},i^*})^2 \quad (38)$$

and J_T is the terminal cost:

$$J_T(\tilde{c}_{L,k+N|k}, \tilde{T}_{L,k+N|k}, e_{\tilde{c},k+N|k}, e_{\tilde{T},k+N|k}) =$$
$$p_1(\tilde{c}_{L,k+N|k} - \tilde{c}_{L,\text{sp},i^*})^2 + p_2(\tilde{T}_{L,k+N|k} - \tilde{T}_{L,\text{sp},i^*})^2 + p_3 e_{\tilde{c},k+N|k}^2 + p_4 e_{\tilde{T},k+N|k}^2$$
$$(39)$$

In (38), (39), $q_i > 0$, $i = 1, \ldots, 5$ and $p_i > 0$, $i = 1, \ldots, 4$ are weighting coefficients. The optimal solution to problem P2 is denoted $U^* = [u_k^*, u_{k+1}^*, \ldots, u_{k+N-1}^*]$ and the control input is chosen according to the receding horizon policy $u(k) = u_k^*$.

The scheme of the adaptive optimal control system for the tubular reactor is shown in Fig. 4. It is a supervisory control system, where the optimal set-point values $\tilde{c}_{L,\text{sp},i^*}$, $\tilde{T}_{L,\text{sp},i^*}$ and the respective steady-state value u_{steady,i^*} are computed in advance by solving problem P1 and stored in the table automata. This automata contains also the multi-models (15)–(16). The scheme includes also a block for identification of the value $\tilde{T}_{f,i^*} \in \mathcal{T}_f$ that is most close to the current measured value $\tilde{T}_f$ of the feed temperature.

4.2 Simulation Results

The adaptive MPC controller based on GP multi-models of the tubular reactor has prediction horizon $N = 20$ and weighting coefficients $q_1 = q_2 = 300$, $q_3 = q_4 = 0.2$, $q_5 = 1 \cdot 10^{-6}$, $p_i = q_i$, $i = 1, \ldots, 4$ in the cost functions (38), (39). The sampling time is 0.1 [s] and the optimization variables in problem P2 are 20 (the values of the coolant flowrate along the horizon). The performance of the MPC controller in closed-loop with

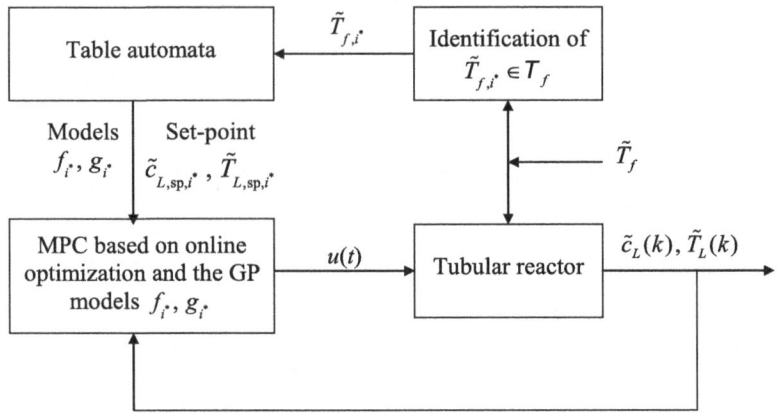

Fig. 4. Adaptive optimal control system based on multi-models and MPC.

the first-principles model (4)–(7) of the tubular reactor is studied. The following scenario is considered. At the discrete time instant 70, the feed temperature changes its value from $\tilde{T}_f = 3$ to $\tilde{T}_f = 3.1$. The control input trajectory generated by the adaptive MPC is shown in Fig. 5 and the corresponding spatiotemporal trajectories of concentration and temperature are given in Fig. 6.

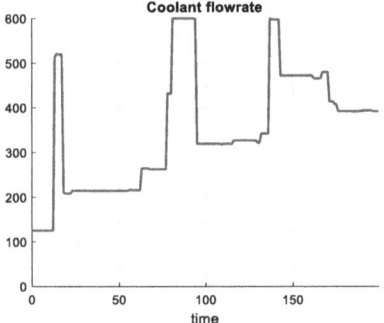

Fig. 5. Control input trajectory.

In Fig. 7, the time trajectories of output concentration and temperature are depicted. From Fig. 7 it can be observed that the dynamics of the output concentration and temperature correspond to a time-delay system. Also, the overall response includes two transients: the first one (more pronounced) is due to the fact that the initial values of the output concentration and temperature are far from the set-point values corresponding to $\tilde{T}_f = 3$ and the second one is related to the change of the feed temperature to a new value $\tilde{T}_f = 3.1$.

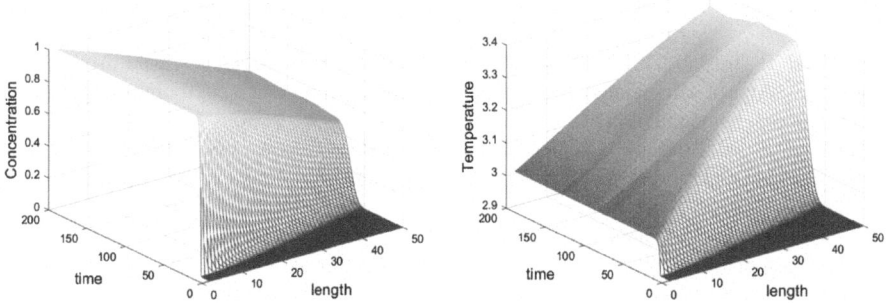

Fig. 6. Spatiotemporal trajectories of concentration (left) and temperature (right).

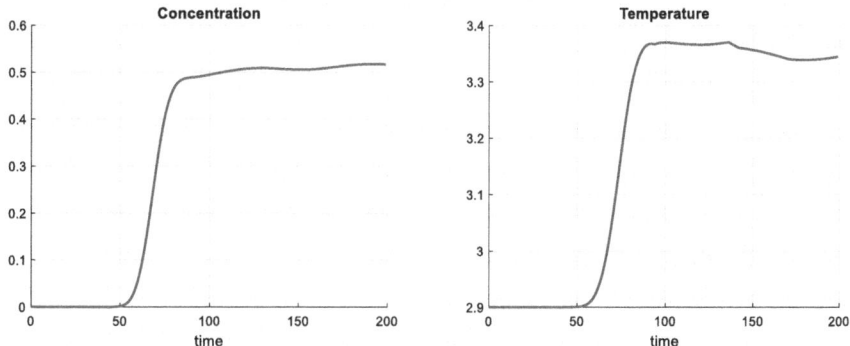

Fig. 7. Time trajectories of output concentration (left) and output temperature (right).

5 Conclusions

In this paper, a novel study on the modeling and adaptive optimal control of a tubular reactor is made by using Gaussian processes. The GP modeling approach is applied to obtain a set of NARX models of the output concentration and temperature of the reactor for several values of the feed temperature. The identified models are then used to design an adaptive model predictive controller to achieve optimal performance of the reactor despite of the changes in the feed temperature. The adaptive system is implemented as table automata that stores the preliminary computed optimal set-point values of the concentration and temperature for different values of the feed temperature. Then, the optimal regulation of the output variables to these set-point values is achieved by the MPC. The simulation experiments show that the adaptive MPC controller brings the output concentration and temperature of the reactor to the new optimal set-point values when the feed temperature changes.

References

1. Bradford, E., Imsland, L., Zhang, D., del Rio Chanona, E.A.: Stochastic data-driven model predictive control using Gaussian processes. Comput. Chem. Eng, (2020). https://doi.org/10.1016/j.compchemeng.2020.106844
2. Grancharova, A., Kocijan, J., Johansen, T.A.: Explicit stochastic nonlinear predictive control based on Gaussian process models. In: European Control Conference, Kos, Greece, 2–5 July 2007, pp. 2340–2347 (2007)
3. Grancharova, A., Kocijan, J., Johansen, T.A.: Explicit stochastic predictive control of combustion plants based on Gaussian process models. Automatica **44**, 1621–1631 (2008)
4. Grancharova, A., Valkova, I., Hvala, N., Kocijan, J.: Distributed predictive control based on Gaussian process models. Automatica **149**, 110807 (2023)
5. Hewing, L., Kabzan, J., Zeilinger, M.N.: Cautious model predictive control using Gaussian process regression. IEEE Trans. Control Syst. Technol. **28**, 2736–2743 (2019)
6. Hicks, G.A., Ray, W.H.: Approximation methods for optimal control synthesis. Can. J. Chem. Eng. **49**, 522–528 (1971)
7. Himmel, A., Matschek, J., Kok, R., Morabito, B., Nguyen, H.H., Findeisen, R.: Machine learning for control of (bio) chemical manufacturing systems. In: Soroush, M., Braatz, R.D. (eds.) Artificial Intelligence in Manufacturing, pp. 181–240. Academic Press (2024)
8. Kocijan, J., Girard, A., Banko, B., Murray-Smith, R.: Dynamic systems identification with Gaussian processes. Math. Comput. Model. Dyn. Syst. **11**, 411–424 (2005)
9. Murray-Smith, R., Shorten, R. (eds.): Switching and Learning in Feedback Systems. LNCS, vol. 3355. Springer, Heidelberg (2005). https://doi.org/10.1007/b105497
10. Kocijan, J.: Modelling and Control of Dynamic Systems Using Gaussian Process Models. Advances in Industrial Control, Springer (2016)
11. Köhler, R., et al.: Methods of lines within the simulation environment DIVA© for chemical processes. In: Vande Wouwer, A., Saucez, P., Schiesser, W.E. (eds.) Adaptive Method of Lines, pp. 371–406. CRC Press, New York, NY, USA (2001)
12. Lahr, A., Zanelli, A., Carron, A., Zeilinger, M.N.: Zero-order optimization for Gaussian process-based model predictive control. Euro. J. Control, 100862 (2023)
13. Likar, B., Kocijan, J.: Predictive control of a gas-liquid separation plant based on a Gaussian process model. Comput. Chem. Eng. **31**, 142–152 (2007)
14. Murray-Smith, R., Sbarbaro, D., Rasmussen, C.E., Girard, A.: Adaptive, cautious, predictive control with Gaussian process priors. In: 13-th IFAC Symposium on System Identification, Rotterdam (2003)
15. Nguyen-Tuong, D., Seeger, M., Peters, J.: Computed torque control with nonparametric regression models. In: 2008 American Control Conference, Seattle, Washington (2008)
16. Nguyen-Tuong, D., Seeger, M., Peters, J.: Real-time local GP model learning. In: Sigaud, O., Peters, J. (eds.) From Motor Learning to Interaction Learning in Robots. Studies in Computational Intelligence, vol. 264. Springer, Heidelberg (2010). https://doi.org/10.1007/978-3-642-05181-4_9
17. Petelin, D., Kocijan, J.: Control system with evolving Gaussian process model. In: IEEE Symposium Series on Computational Intelligence, SSCI 2011. IEEE, Paris (2011)
18. Rasmussen, C.E., Williams, C.K.I.: Gaussian Processes for Machine Learning. The MIT Press, Cambridge (2006)
19. Rose, A., Pfefferkorn, M., Nguyen, H.H., Findeisen, R.: Learning a Gaussian process approximation of a model predictive controller with guarantees. In: 62nd IEEE Conference on Decision and Control, pp. 4094–4099 (2023)
20. Sbarbaro, D., Murray-Smith, R.: Self-tuning control of nonlinear systems using Gaussian process prior models. In: Shorten, R., Murray-Smith, R. (eds.) Switching and Learning in Feedback Systems, LNCS, vol. 3355, pp. 140–157. Springer, Heidelberg, Germany (2005)

21. Solak, E., Murray-Smith, R., Leithead, W.E., Leith, D.J., Rasmussen, C.E.: Derivative observations in Gaussian process models of dynamic systems. In: NIPS 15. MIT Press, Vancouver (2003)
22. Wouwer, A.V., Saucez. P., Vilas, C.: Simulation of ODE/PDE models with MATLAB, OCTAVE and SCILAB. In: Scientific and Engineering Applications, pp. 125–202. Springer, Cham (2014).

Converging Dimensions: Information Extraction and Summarization Through Multisource, Multimodal, and Multilingual Fusion

Pranav Janjani[✉], Mayank Palan, Sarvesh Shirude, Ninad Shegokar, and Faruk Kazi

Centre of Excellence in Complex and Nonlinear Dynamical Systems,
VJTI, Mumbai, India
prjanjani_b21@ce.vjti.ac.in, mbpalan_b22@it.vjti.ac.in,
{seshirude_b20,fskazi}@el.vjti.ac.in, nsshegokar_b23@ee.vjti.ac.in

Abstract. Traditional summary generation approaches are limited by their reliance on isolated sources of data, restraining the quantity and quality of information gathered. This introduces the possibility of falsified content and provides limited support for multilingual and multimodal data. This paper presents a novel approach to summarization that tackles such challenges by utilizing the strength of multiple sources to deliver a more exhaustive and informative understanding of intricate topics. It progresses beyond conventional, unimodal sources such as text documents, integrating a diverse range of data, including YouTube playlists, pre-prints, and Wikipedia pages. The aforementioned multimodal sources are converted into a unified textual representation, enabling a holistic analysis. This multifaceted approach empowers us to extract pertinent information from a wider array of sources. The primary tenet of this approach is to maximize information gain, maintain a high level of informativeness while minimizing information overlap, and encourage the generation of highly coherent summaries.

Keywords: Information Retrieval · Information Fusion · Multi-Source Summarization · Multimodal Summarization

1 Introduction

Within the contemporary paradigm of omnipresent information access, where knowledge dissemination transcends traditional boundaries of format and language, the ability to efficiently extract and synthesize significant volumes of data becomes paramount. Methods have been implemented for textual summarization on PDFs from sources like educational documents and scientific research papers [1,2]. For research, there are significant amounts of unused information in the form of a large number of scholarly articles on websites like arXiv [3,4]. An intelligent paper search module is integrated that works just like a seasoned navigator, which searches for appropriate research papers according to the query entered by the user. This focused approach mitigates the need to go through a plethora of irrelevant information. Once research paper harvesting is complete, summary methodology is employed to collate the retrieved papers into short, clear, and informative summaries [5]. This way, researchers and students can quickly investigate the latest developments in their area of interest without expending too much time and effort.

YouTube is another source of information where the process employed is more advanced than just textual analysis, it also generates transcripts from multilingual videos [6,7]. Advance methods in extracting frames to capture the visual narrative embedded in the film are employed [8]. By dissecting what is being said, one can better understand the topic. Therefore, the keyframes and generated transcripts are used to produce the summary. To further improve the context of the information retrieved, summaries from reliable knowledge base such as Google, DuckDuckGo, and Wikipedia are concatenated [9,10]. These resources are vast libraries of filtered, verified human knowledge and provide essential background information. By integrating information from the sources to which it is connected, the system thus manages a more profound understanding of the underlying relationships between concepts and entities. A summary of a scientific result is extracted from these vast information banks that not only describes the result but also embeds it in the appropriate historical context and the framework of scientific ideas. This well-defined structure allows a deep understanding of the content, enabling the easy integration of information from several modalities and languages into the document [11].

This paper presents a methodology for multi-source summarization, capable of gleaning nuanced details and insights from a confluence of heterogeneous data streams - multilingual, multimodal, and encompassing the full spectrum of textual, visual, and conceptual information. This holistic approach fosters a cohesive understanding by leveraging the complementary strengths of diverse information sources, ensuring a comprehensive and multifaceted representation of the underlying knowledge. This will enhance user comprehension and make navigating the ever-increasing knowledge landscape smoother by combining the power of several sources into one joint knowledge base.

The proposed multifaceted approach should serve a dual purpose: it not only mitigates redundancy within the extracted data but also promotes the inclusion of diverse and potentially conflicting perspectives. This comprehensive methodology should enhance the overall quality of the data by optimizing its relevance and breadth. Minimizing repetitive information, ensures the capture of a broader spectrum of perspectives and insights, thereby enriching the dataset with a nuanced and multifarious understanding of the subject matter.

2 Related Work

The wealth of multimodal and multilingual information demands valuable ways of summarizing and synthesizing information with minimal overlap. Recent works have released a multilingual abstractive summarization test set using a large dataset with multiple languages, human summaries, and pairs of the source document and its summary in different languages [12,13]. This milestone work provides evidence of an opportunity for automatic systems to synthesize culturally appropriate and coherent summaries. Based on such principles, video summarization frameworks have appeared with the addition of transcription and thematic analysis of video content. For instance, a documentary synopsis is supposed to include important visuals and summarize the major events and characters covered [14]. All these go a step further than merely summarizing text, considering audiovisual aspects in making a holistic narrative [6,15]. Providing summaries of videos that focus on the key visual and thematic elements have also been outlined [16].

Some techniques emphasized retrieval-based mechanisms to improve the relevance and informativeness characteristics of summaries [17]. Summarizing research papers calls for specialized models and domain adaptation techniques to deal with a broad scope of domain-specific challenges, including scientific terms and complex syntactic structures. An attractive approach for summarizing scientific documents through joint fact detection in citations is proposed which identifies noun phrases from citation sentences that frequently co-occur. It represent common facts discussed from different perspectives. These facts could then be compiled using this multi-document summarization system into a comprehensive summary [18]. Another literature investigates multi-document summarization techniques that evaluate various models, including graph-based, neural network-based, and hybrid approaches, proposing methods to effectively synthesize information from multiple papers [19].

The prevailing information extraction and summarization methodologies are predominantly characterized by their singular source dependence and a dearth of multi-modality. This constraint restricts the potential for optimal knowledge acquisition. This singular source dependence often leads to redundancy within extracted information and hinders the capture of diverse perspectives. To achieve a more comprehensive understanding of a subject, it is imperative to leverage information from a multiplicity of sources. To address these limitations, research efforts should be directed towards the development of robust multi-source information extraction and summarization techniques.

3 Methodology

This paper proposes a multisource, multimodal, and multilingual information extraction system, the first of its kind to capture the most important and diverse information to reduce hallucinations and increase the quality of summary generation. Figure 1, provides a comprehensive overview of the process. The functions and methods are categorized into these parts:

1. Information Conversion
2. Information Search & Retrieval
3. Information Convergence

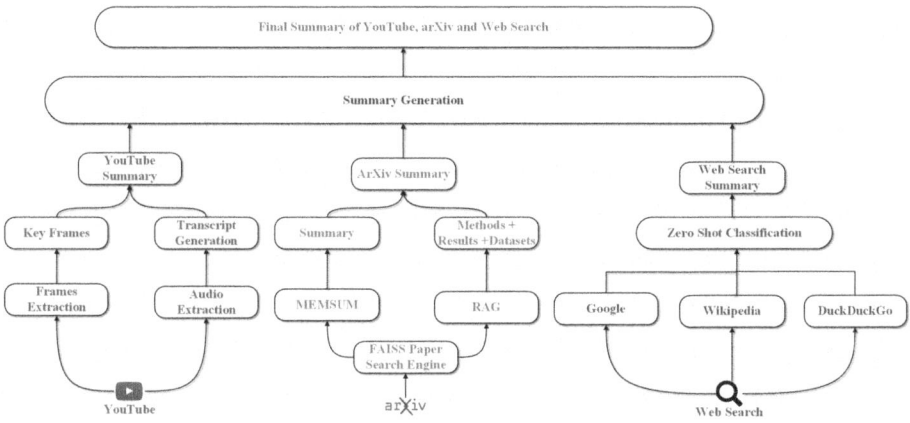

Fig. 1. Methodology

Three multimodal sources- YouTube Playlists, arXiv Papers and Web Search are considered. Each is passed through the mentioned processes to generate a superior system capable of providing robust information related to any subject matter.

3.1 YouTube Playlists: Multilingual and Multimodal Approach

Within the dynamic domain of information extraction, the primary objective revolves around the harvesting of valuable knowledge from disparate sources, YouTube playlists emerge as an intriguing and relatively unexplored research frontier. By incorporating information gleaned from playlists in conjunction with established textual sources, a system can cultivate a more comprehensive and richly multifaceted comprehension of the target domain

Information Conversion: YouTube encompasses multimodal information- audio and video each processed separately and merged to generate an optimal information representation of the query presented by the user.

(i) Audio Extraction: The audio stream gets extracted using the YouTube API and the audio is subsequently converted to text along with its timestamp using Open AI's robust speech recognition model Whisper [20]. It can identify the language of the video and transcribe the audio to a desired language. An alternate method of using YouTube's Closed Caption generation was explored and provided poor results due to the lesser accuracy of the captions generated and limited multilingualism. The mentioned support is provided only for a smaller percentage of the videos so it's not optimal, adaptable or scalable.

(ii) Video Frames Extraction: The process begins with extracting frames from the video, calculating frame differences, identifying local maxima, and selecting high-quality frames through clustering [21,22]. The resulting keyframes are processed using OCR to extract the text for a comprehensive inclusion of information. Figure 2 illustrates frame extraction.

To identify frames that exhibit significant changes, we calculate the sum of absolute differences (SAD) between consecutive frames which involves Conversion to HSV where each frame is converted to the HSV color space and a change between frames is identified by calculating a difference.

A custom Frame class stores each frame along with its computed SAD value. These values help identify frames with substantial content changes. To extract frames that are significantly different from their neighbors, we use local maxima detection. We start by smoothing the frame difference values using a window unction subsequently the local maxima are identified.

Frames corresponding to these local maxima are considered candidate keyframes, representing moments of significant visual change in the video. Once candidate keyframes are extracted, they undergo further evaluation based on brightness, entropy, and sharpness to ensure high visual quality.

Frames with optimal brightness and contrast (entropy) are filtered for further analysis. To ensure diversity among the selected keyframes, clustering techniques are employed. First, each candidate keyframe is converted to grayscale and resized to a standard dimension. The DCT is then applied to capture frequency components, which are used as feature vectors. The feature vectors are clustered using HDBSCAN, which groups frames into clusters based on similarity. Within each cluster, the frame with the highest Laplacian variance is

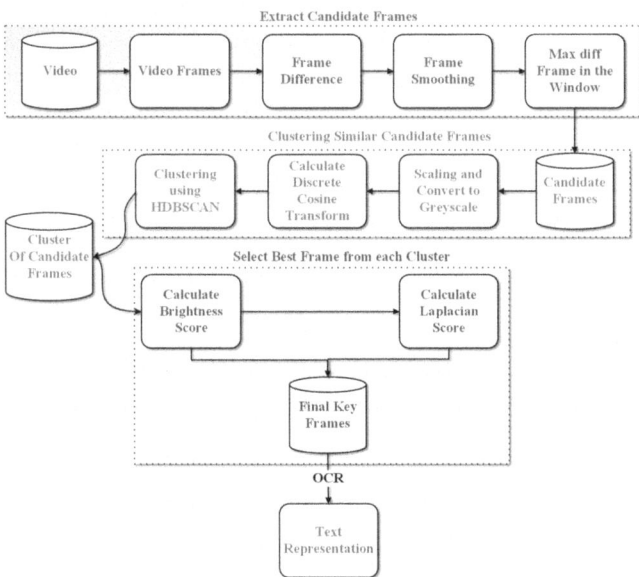

Fig. 2. Key frames and Information Extraction

selected as the best representative frame. This step ensures that the final set of key frames is sharp and visually distinct.

After the key frames are isolated and synchronized with corresponding timestamps, the visual data undergoes text recognition and extraction. Google's Generative Pre-trained Transformer, also known as Gemini, is utilized for this purpose of Optical Character Recognition. The resulting time-stamped textual elements, gleaned from the processed video frames, are subsequently subjected to a data fusion process. This entails the seamless integration of the extracted time-stamped text with the existing, synchronized transcript. This enriched dataset, encompassing both visual and textual information with precise temporal alignment, serves as the foundation for subsequent information convergence process.

Information Search and Retrieval: Open-sourced Playlist search libraries are utilized to facilitate real-time information retrieval. The limitations of static datasets are eschewed in favor of a dynamic approach that incorporates the ever-evolving corpus of information on YouTube [23]. This approach facilitates the retrieval of a vast volume of playlists, which are then be subjected to an optimal information conversion process. By simple entry of keywords as queries, the libraries retrieve the relevant playlists.

Information Convergence: To efficiently extract key information from video playlists, a synopsis generation paradigm is employed. This approach entails processing each video within the playlist to produce a concise summary that retains crucial details while eliminating redundant content. To tailor summaries to the specific content domain of the playlist, the LLaMA3 70b model's power is coupled with meticulously crafted prompt engineering techniques. Through this synergistic approach, it is ensured that the generated summaries accurately capture the diverse content of each video and the overarching narrative of the playlist.

3.2 arXiv Paper Search and Summarization

The inclusion of arXiv and research papers is quite beneficial and indispensable. They offer a unique window into the frontiers of scientific knowledge, providing access to cutting-edge discoveries, detailed exploration of all related topics, and a structured format that facilitates accurate information extraction. By incorporating these valuable resources, the system can gain a deeper understanding of the information landscape and extract more comprehensive, nuanced, and insightful knowledge.

Information Conversion: The arXiv dataset is a repository of 1.7 million articles [24], with relevant features such as article id, titles, abstract, full text PDFs, and more. The title and abstract for each paper were combined and converted to vector embedding using sentence transformers 'all-MiniLM-L6-v2' model. It maps sentences & paragraphs to a 384-dimensional dense vector space and can be used for various natural language tasks like semantic search [25].

Information Search and Retrieval: Retrieval Augmented Generation (RAG) system using Facebook AI Similarity Search (FAISS) was built to search for research papers in the arXiv dataset based on the user query and retrieves the top 'n' papers [26]. These embedded vectors are stored in the RAG System using a FAISS index, which is an open-source library designed to quickly search for embeddings in large datasets and provides scalable similarity search functions [27].

Upon submission of a user search query, the query text will be vectorized. These query vectors will then be compared to the vectors stored in the FAISS index. Further re-ranking is using the 'cross-encoder/ms-marco-MiniLM-L-6-v2' encoder model to improve vector search results as it compares text instead of vectors. Overall, it reorders the text search result based on the search query by assigning a score to the text result. The re-ranked search results are returned which consists of paper title, abstract, its categories, and a link to the PDF. Text extraction techniques are employed to convert the document, located at a specific URL, into a machine-readable format.

Information Convergence: Subsequently, the summarization engine consists of two distinct levels to ensure the capture of all critical information. The first level focuses on the faithful extraction of statistical and mathematical expressions within the paper while the second level delves into the background knowledge presented and the novel findings introduced by the research, guaranteeing an inclusive and informative summary. This ensures the inclusion of crucial statistical and mathematical equations, while simultaneously retaining the significance of background knowledge and novel findings presented within the research paper.

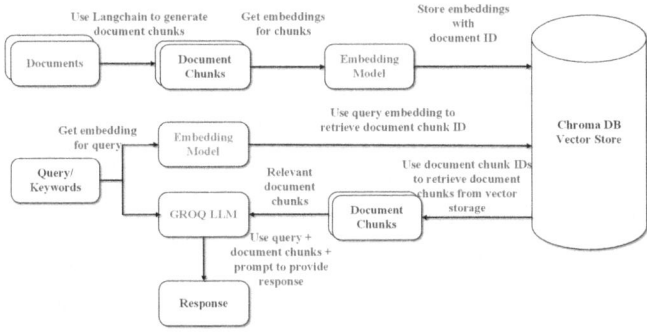

Fig. 3. RAG Chain Pipeline

(i) MeMSum Pipeline: The extracted text is summarized using Multi-step Episodic Markov decision process extractive Summarizer (MeMSum). MemSum's 'mesum-arXiv-summarization' pre-trained model is used to summarize the research papers [28].
(ii) RAG Chain Pipeline: The research paper's information is segmented into manageable chunks using Langchain. Each chunk is then transformed into a numerical representation, or embedding, via an embedding model, which is 'BAAI/bge-small-en-v1.5' capturing semantic nuances necessary for efficient retrieval [29]. These embeddings, alongside document IDs, are stored in a Chroma DB Vector Store, enabling quick, similarity-based retrieval. This storage mechanism ensures that the document chunks can be easily accessed and utilized in subsequent stages of the pipeline. Figure 3 illustrated the mechanisms of the RAG chain pipeline.

Upon receiving a query, an embedding model generates a corresponding query embedding, which is utilized to fetch relevant document chunks from the Chroma DB Vector Store. These chunks, containing pertinent information, are fed into a GROQ LLM, which processes the query and contextually integrates the document chunks. The LLM is prompted to retrieve and synthesize information specifically about the methodology, datasets, and results of the research papers. This approach maximizes the utility of large language models by augmenting them with specific, contextual data retrieved efficiently through embeddings and vector stores, ensuring the responses are accurate, relevant, and comprehensive.

Following the independent processing by the MeMSum and RAG Chain pipelines, their outputs are synergistically merged to create a holistic textual representation. The resulting comprehensive nature of this information source facilitates the efficacious dissemination of knowledge. This ensures that users, particularly those engaged in academic pursuits (i.e., students and researchers), possess a contemporaneous understanding of the latest developments within their respective fields.

3.3 Web Search Engine

To mitigate the risk of data obsolescence and guarantee exhaustive information retrieval from a diverse range of sources, a multifaceted web search strategy is employed. This strategy leverages the capabilities of prominent search engines, namely Wikipedia, DuckDuckGo, and Google.

Information Search and Retrieval: The aforementioned search engines function as established web crawlers, systematically traversing the web to index content provided as "llama-index" agents. Extracted webpages undergo a rigorous topic classification process using a large language model (LLM) based Zero-Shot classification technique [30]. This classification ensures that only those webpages demonstrably aligned with the research query's thematic domain are retained for subsequent analysis.

Information Convergence: This involves synthesizing a vast array of pertinent data retrieved from numerous web search engines. Key insights are distilled from this extensive collection, significantly reducing redundant information and ensuring the content remains current and relevant. This is accomplished by harnessing the formidable capabilities of the LLaMA3 70b model, expertly combined with meticulously crafted prompt engineering techniques [31].

3.4 Final Summary Generation

Following the derivation of succinct representations from each source, MeMSum in a multi-source fashion is used to strategically mitigate the redundancy and overlap of information across the summaries. This approach ensures the inclusion of all salient and distinct viewpoints, fostering the generation of a comprehensive information extraction summary that encompasses multi-source, multi-modal, and multilingual content.

4 Metrics

The following robust metrics are utilized to rigorously evaluate the efficacy of our information extraction and summarization system:

4.1 Entropy

Entropy in textual summaries measures the average unpredictability or information richness of content. This dimension shows the coverage of vocabulary and topics gained from the text summary, i.e., overall diversity and coverage. Higher values for entropy suggests an extensive spread of different words and concepts and leading towards an extremely informative and diverse summary. In contrast, lower entropy implies a predictable structure with less variety of unique terms, suggesting a narrower focus or repetitive content [32].

$$H(X) = -\sum_{i=1}^{|X|} p(x_i) \log_2(p(x_i)) \tag{1}$$

where, $H(X)$ represents the entropy of the document X, $|X|$ represents the total number of words in the document, and $p(x_i)$ represents the probability of the i-th word appearing in the document.

4.2 KL Divergence

Kullback-Leibler (KL) Divergence measures the difference between two probability distributions of summary content. It quantifies how much more information or content one summary adds compared with another. Small KL Divergence indicates similar content distributions for summaries, while larger values suggest more divergence and unique information-diverse content [33].

$$D_{KL}(P||Q) = \sum_{i=1}^{|X|} p(x_i) \log_2\left(\frac{p(x_i)}{q(x_i)}\right) \qquad (2)$$

where, $D_{KL}(P||Q)$ represents the KL divergence between two probability distributions P and Q which represent the word distributions of two different documents or a document and its language model, $p(x_i)$ represents the probability of the i-th word under distribution P and $q(x_i)$ represents the probability of the i-th word under distribution Q.

4.3 Redundancy Score

If the summary has a low redundancy score, it brings novel perspectives and information not found in other summaries; hence, it becomes invaluable for different views or information. Higher redundancy scores reflect more overlap or repetition in content with other summaries, whereas lower scores show less unique contribution in information.

4.4 Average Coherence

It measures semantic consistency and flow of the summary by checking inter-sentential similarity between sentences or paragraphs; that is, how good ideas are connected and presented within the summary. Higher average coherence scores mean a summary is more well-structured and coherent, with smoother transitions among points. Lower average scores may indicate disjointed or less coherent content, where ideas may not flow as logically or seamlessly [34].

$AverageCoherence =$

$$\frac{1}{n-w+1} \sum i = w^n \frac{\left(\frac{1}{w}\sum_{j=i-w}^{i-1} E_j\right) \cdot E_i}{\left\|\frac{1}{w}\sum_{j=i-w}^{i-1} E_j\right\| \|E_i\|} \qquad (3)$$

4.5 Evaluating Document Coherence Modeling

This paper examines how well different language models can evaluate document coherence. It delves into the methods used to measure semantic consistency and flow within summaries, specifically focusing on inter-sentential similarity and the logical progression of ideas [34].

4.6 Type-Token Ratio (TTR)

The Type-Token Ratio (TTR) is a measure of lexical diversity in a given text. It is defined as the ratio of the number of unique words (types) to the total number of words (tokens) in the text [35].

$$TTR = \frac{V}{N} \qquad (4)$$

where, V represents the number of unique words (types) in the text and N is the total number of words (tokens) in the text.

4.7 ROUGE Scores

Recall-Oriented Understudy for Gisting Evaluation (ROUGE) scores compare summaries by measuring their efficiency at capturing information critical in reference texts. The same n-grams (words, phrases) overlapping between a generated summary and reference text show the statistics of precision, recall, and F1 measures for these.

High ROUGE scores reflect high agreement and overlap between a generated summary and its reference, being more exact or complete in that sense. Low scores indicates gaps or incongruities in the information or misunderstanding of the original content of the summary [36].

4.8 Results Discussion

The methodology presented in the paper has been extensively evaluated over a plethora of examples. Three samples - 'Deep Learning', 'Statistics' and 'Quantum Physics' are presented below.

Table 1. Comparison of Coherence Score

Keyword	Coherence Score
Deep Learning	0.470
Stats	0.462
Quantum	0.470

Table 1 depicts that the final summaries are coherent, they follow a logical flow and allow the reader to navigate the information extracted sensibly and it ensures the abruption caused due to multiple sources information integration is minimal.

Tables 2, 3, and 4 contain metrics used to evaluate the quality of the final summary compared to summaries from individual sources. The KL Divergence metric suggests that the divergence between individual sources as evident is quite high, which supports our claim that the information brought in by different sources is unique in nature. Relatively, the KL divergence between Final Summary and different sources is less which shows that our final summary integrates information from different sources properly. This integration results in a summary that is diversified and rich in information content, as indicated by high entropy. It utilizes an enriched vocabulary, reflected by the type token ratio, and effective in expressing information, as shown by the low redundancy score. Consequently, information from different sources makes the description of the

Table 2. Metrics Evaluation for 'Deep Learning' sample

Metric	Final Summary	arXiv	Web Search	YouTube
KL Divergence (vs arXiv)	0.465	-	4.791	4.431
KL Divergence (vs Web Search)	2.360	5.424	-	5.057
KL Divergence (vs YouTube)	1.404	4.861	4.808	-
Entropy	8.113	7.540	7.993	7.739
Semantic Density	0.186	0.218	0.254	0.134
Redundancy Score	0.084	0.453	0.578	0.235

Table 3. Metrics Evaluation for 'Statistics' sample

Metric	Final Summary	arXiv	Web Search	YouTube
KL Divergence (vs arXiv)	1.824	-	4.453	3.599
KL Divergence (vs Web Search)	0.442	4.867	-	2.636
KL Divergence (vs YouTube)	0.925	5.141	4.118	-
Entropy	8.491	7.998	7.816	8.576
Semantic Density	0.133	0.289	0.207	0.109
Redundancy Score	0.055	0.810	0.521	0.088

Table 4. Metrics Evaluation for 'Quantum Physics' sample

Metric	Final Summary	arXiv	Web Search	YouTube
KL Divergence (vs arXiv)	2.250	-	5.252	4.350
KL Divergence (vs Web Search)	0.449	4.900	-	2.524
KL Divergence (vs YouTube)	0.704	4.531	3.442	-
Entropy	8.348	8.072	7.925	8.203
Semantic Density	0.158	0.317	0.255	0.148
Redundancy Score	0.038	0.693	0.387	0.114

Table 5. ROUGE Score metric for 'Deep Learning' sample

Category	Rouge-1			Rouge-2			Rouge-L		
	Recall	Precision	F1	Recall	Precision	F1	Recall	Precision	F1
Final Summary vs arXiv	0.955	0.446	0.608	0.941	0.373	0.534	0.954	0.446	0.608
Final Summary vs Web Search	0.747	0.381	0.505	0.702	0.273	0.394	0.745	0.380	0.504
Final Summary vs YouTube	0.553	0.505	0.528	0.411	0.442	0.426	0.545	0.498	0.520
arXiv vs Web Search	0.233	0.255	0.243	0.062	0.061	0.062	0.208	0.227	0.217
arXiv vs YouTube	0.177	0.345	0.234	0.046	0.125	0.068	0.166	0.324	0.219
Web Search vs YouTube	0.161	0.288	0.206	0.031	0.086	0.046	0.145	0.259	0.186

Table 6. ROUGE Score metric for 'Statistics' sample

Category	Rouge-1			Rouge-2			Rouge-L		
	Recall	Precision	F1	Recall	Precision	F1	Recall	Precision	F1
Final Summary vs arXiv	0.714	0.204	0.317	0.553	0.107	0.179	0.693	0.198	0.308
Final Summary vs Web Search	1.000	0.380	0.551	1.000	0.293	0.453	1.000	0.380	0.551
Final Summary vs YouTube	0.622	0.719	0.667	0.537	0.682	0.601	0.617	0.713	0.661
arXiv vs Web Search	0.216	0.288	0.247	0.051	0.078	0.062	0.195	0.259	0.222
arXiv vs YouTube	0.108	0.438	0.174	0.026	0.168	0.044	0.101	0.408	0.161
Web Search vs YouTube	0.150	0.456	0.226	0.040	0.175	0.066	0.143	0.434	0.215

Table 7. ROUGE Score metric for 'Quantum Physics' sample

Category	Rouge-1			Rouge-2			Rouge-L		
	Recall	Precision	F1	Recall	Precision	F1	Recall	Precision	F1
Final Summary vs arXiv	0.696	0.276	0.395	0.577	0.152	0.240	0.683	0.271	0.388
Final Summary vs Web Search	0.968	0.397	0.563	0.956	0.306	0.464	0.968	0.397	0.563
Final Summary vs YouTube	0.774	0.693	0.731	0.711	0.651	0.679	0.766	0.686	0.724
arXiv vs Web Search	0.232	0.241	0.236	0.058	0.070	0.064	0.206	0.213	0.210
arXiv vs YouTube	0.171	0.386	0.237	0.040	0.138	0.062	0.156	0.353	0.217
Web Search vs YouTube	0.229	0.499	0.313	0.078	0.224	0.116	0.209	0.456	0.286

topic more detailed and balanced compared to what would have been achieved in any one source.

From Table 5, 6 and 7 it is evident that high recall scores in comparisons with individual sources indicate that the final summary contains most relevant content from the individual sources. The moderate values of the precision scores prove that the final summary has more context or information than the individual sources. The F1 scores of the final summary are consistent, indicating a balance between recall and precision. The shallow ROUGE scores of the three different sources, arXiv, Wikipedia, and YouTube, document their particular contribution to the presented work. The high distinctiveness value in the ROUGE scores points to the worth of combining these multiple sources, particularly in making a final summary of worth both rich and detailed. In general, the hypothesis that the given topic is fully covered in this multi-channel, multimodal and multilingual setting, taking advantage of idiosyncrasy offered by each source, gets backed up by the ROUGE scores.

5 Conclusion

The polyvalent approach, defined by the removal of redundant data and the strategic inclusion of opposing viewpoints, significantly improved the thematic relevance and scope. A comparative evaluation of metrics from various information sources, including entropy, KL divergence, type token ratio, and redundancy scores, demonstrated the effectiveness of this methodology. Notably, the generated summaries had a low level of redundancy combined with an optimal type token ratio. Furthermore, the use of the ROUGE metric demonstrated the intricate relationship between textual similarity and coverage in both the final product and the source materials. This emphasizes the significance of incorporating diverse perspectives in order to acquire a comprehensive understanding. Furthermore, the proposed framework exhibited exceptional coherence, bolstering its capacity to safeguard thematic pertinence and internal consistency. The metrics and comparative analyses presented throughout this work provide unequivocal support for the effectiveness of this strategy, solidifying its potential to significantly impact the knowledge foundation and analytical resilience of technical research endeavors in summary generation.

Acknowledgments. Authors acknowledge the support of the Centre of Excellence (CoE) in Complex & Nonlinear Dynamical Systems (CNDS) under TEQIP-III funding.

References

1. Allahyari, M., et al.: Text summarization techniques: a brief survey. arXiv preprint arXiv:1707.02268 (2017)
2. Liu, Y., Lapata, M.: Text summarization with pretrained encoders (2019)
3. Jiang, F., Wang, K., Li, H.: Bridging research and readers: a multi-modal automated academic papers interpretation system (2024)
4. Ibrahim Altmami, N., El Bachir Menai, M.: Automatic summarization of scientific articles: a survey. J. King Saud Univ. Comput. Inf. Sci. **34**(4), 1011–1028 (2022)
5. An, C., Zhong, M., Geng, Z., Yang, J., Qiu, X.: Retrievalsum: a retrieval enhanced framework for abstractive summarization (2021)
6. Lin, J., et al.: Videoxum: cross-modal visual and textual summarization of videos (2024)
7. Sun, H., Zhu, X., Zhou, C.: Deep reinforcement learning for video summarization with semantic reward. In: 2022 IEEE 22nd International Conference on Software Quality, Reliability, and Security Companion (QRS-C), pp. 754–755 (2022)
8. Ciocca, G., Schettini, R.: Erratum to: an innovative algorithm for key frame extraction in video summarization. J. Real-Time Image Process. **1**, 69–88 (2006)
9. Hingu, D., Shah, D., Udmale, S.: Automatic text summarization of Wikipedia articles view document (2015)
10. Nemoto, Y., Klyuev, V.: Tool to retrieve less-filtered information from the internet. Information **12**(2), 65 (2021)
11. Yuan, J., Gao, N., Xiang, J., Tu, C., Ge, J.: Knowledge graph embedding with order information of triplets. In: Yang, Q., Zhou, Z.-H., Gong, Z., Zhang, M.-L., Huang, S.-J. (eds.) PAKDD 2019. LNCS (LNAI), vol. 11441, pp. 476–488. Springer, Cham (2019). https://doi.org/10.1007/978-3-030-16142-2_37
12. Cao, Y., Wan, X., Yao, J., Yu, D.: Multisumm: towards a unified model for multilingual abstractive summarization. In: Proceedings of the AAAI Conference on Artificial Intelligence, vol. 34, pp. 11–18 (2020)
13. Hasan, T., et al.: XL-sum: large-scale multilingual abstractive summarization for 44 languages (2021)
14. Otani, M., Nakashima, Y., Rahtu, E., Heikkilä, J., Yokoya, N.: Video summarization using deep semantic features (2016)
15. Zhou, K., Qiao, Y., Xiang, T.: Deep reinforcement learning for unsupervised video summarization with diversity-representativeness reward (2018)

16. Gianluigi, C., Raimondo, S.: An innovative algorithm for key frame extraction in video summarization. J. Real-Time Image Proc. **1**, 69–88 (2006)
17. Liu, S., Wu, J., Bao, J., Wang, W., Hovakimyan, N., Healey, C.G.: Towards a robust retrieval-based summarization system (2024)
18. Chen, J., Zhuge, H.: Summarization of scientific documents by detecting common facts in citations. Future Gener. Comput. Syst. **32**, 246–252 (2014). Special Section: The Management of Cloud Systems, Special Section: Cyber-Physical Society and Special Section: Special Issue on Exploiting Semantic Technologies with Particularization on Linked Data over Grid and Cloud Architectures
19. Agarwal, N., Reddy, R.S., Kiran, G.V.R., Rose, C.: Scisumm: a multi-document summarization system for scientific articles, pp. 115–120 (2011)
20. Radford, A., Kim, J.W., Xu, T., Brockman, G., McLeavey, C., Sutskever, I.: Robust speech recognition via large-scale weak supervision. In: Proceedings of the 40th International Conference on Machine Learning, ICML 2023. JMLR.org (2023)
21. Sadiq, B., Muhammad, B., Abdullahi, M., Onuh, G., Ali, A., Babatunde, A.: Keyframe extraction techniques: a review. ELEKTRIKA - J. Electr. Eng. **19**, 54–60 (2020)
22. Calic, J., Izuierdo, E.: Efficient key-frame extraction and video analysis. In: Proceedings of the International Conference on Information Technology: Coding and Computing, pp. 28–33 (2002)
23. Abu-El-Haija, S., et al.: Youtube-8m: a large-scale video classification benchmark (2016)
24. arXiv.org submitters. arXiv dataset (2024)
25. Wang, W., Wei, F., Dong, L., Bao, H., Yang, N., Zhou, M.: Minilm: deep self-attention distillation for task-agnostic compression of pre-trained transformers. In: Proceedings of the 34th International Conference on Neural Information Processing Systems, NIPS 2020, Red Hook, NY, USA, Curran Associates Inc. (2020)
26. P. Lewis, E. Perez, A. Piktus, F. Petroni, V. Karpukhin, N. Goyal, H. Küttler, M. Lewis, W.-t. Yih, T. Rocktäschel, S. Riedel, and D. Kiela, "Retrieval-augmented generation for knowledge-intensive nlp tasks," in *Proceedings of the 34th International Conference on Neural Information Processing Systems*, NIPS '20, (Red Hook, NY, USA), Curran Associates Inc., 2020
27. Douze, M., et al.: The faiss library (2024)
28. Gu, N., Ash, E., Hahnloser, R.: MemSum: extractive summarization of long documents using multi-step episodic Markov decision processes. In: Muresan, S., Nakov, P., Villavicencio, A. (eds.) Proceedings of the 60th Annual Meeting of the Association for Computational Linguistics (Volume 1: Long Papers), Dublin, Ireland, pp. 6507–6522. Association for Computational Linguistics (2022)
29. Chen, J., Xiao, S., Zhang, P., Luo, K., Lian, D., Liu, Z.: BGE m3-embedding: multi-lingual, multi-functionality, multi-granularity text embeddings through self-knowledge distillation (2024)
30. Yin, W., Hay, J., Roth, D.: Benchmarking zero-shot text classification: datasets, evaluation and entailment approach (2019)
31. AI@Meta. Llama 3 model card (2024)
32. Zhang, X., Hu, D., Li, B., Qin, Y., Li, L.: Content richness evaluation method for abstractive summarization. In: 2023 IEEE 9th International Conference on Cloud Computing and Intelligent Systems (CCIS), pp. 245–249 (2023)
33. Zhang, Y., Liu, W., Chen, Z., Wang, J., Li, K.: On the properties of Kullback-Leibler divergence between multivariate gaussian distributions (2023)
34. Shen, A., Mistica, M., Salehi, B., Li, H., Baldwin, T., Qi, J.: Evaluating document coherence modeling. Trans. Assoc. Comput. Linguist. **9**, 621–640 (2021)

35. Reviriego, P., Conde, J., Merino Gómez, E., Martínez, G., Hernández, J.: Playing with words: comparing the vocabulary and lexical richness of chatgpt and humans (2023)
36. Lin, C.-Y.: ROUGE: a package for automatic evaluation of summaries. In: Text Summarization Branches Out, Barcelona, Spain, pp. 74–81. Association for Computational Linguistics (2004)

Enhancing Question Answering in Lecture Videos with a Multimodal Retrieval-Augmented Generation Framework

Thomas Tanner[1,2](✉)[iD], Andreas Marfurt[2][iD], and Hasan Oğul[1][iD]

[1] Østfold University College, Halden, Norway
hi@thomastanner.io, hasan.ogul@hiof.no
[2] Lucerne University of Applied Sciences and Arts, Rotkreuz, Switzerland
andreas.marfurt@hslu.ch

Abstract. This paper addresses the challenges of searching and extracting information from lecture videos as found on online platforms. With the increasing popularity of educational videos, learners are faced with the growing challenge of video search and finding relevant sequences within videos to answer questions. This research describes the development of a framework that relies on Retrieval-Augmented Generation (RAG) to address this challenge with a focus on the nature of lecture videos, which usually rely heavily on audio data (voice of the lecturer) and image data (lecture slides) that can partially be extracted as text. The resulting artifact allows a user to ask a question based on a corpus of videos and receive an answer based on the videos' contents, along with information regarding the most relevant videos and sequences within the corpus. This research illustrates that a RAG system leveraging textual context from lecture videos is effective at improving question answering accuracy compared to relying solely on a standalone Large Language Model (LLM) without contextual enhancements. Further findings revealed that boolean answer correctness can aid in optimizing hyperparameters, that the independent retrieval of text from audio and visual modalities outperforms more complex strategies, and that retrieval followed by re-ranking outperforms other strategies.

Keywords: Lecture Video Question Answering · Multimodality · Retrieval-Augmented Generation · Large Language Models · Local Computing · Automatic Speech Recognition · Optical Character Recognition

1 Introduction

Lecture videos and other educational videos are increasing in number and popularity. Massive Open Online Courses (MOOCs) are educational courses that are accessible on the Web, usually in the form of lecture videos and corresponding course content [35]. The amount of available MOOCs has been increasing steadily, from fewer than 1000 in 2012 to more than 19 thousand in 2021 [7]. In 2023, Class Central, a website that allows searching for courses on various platforms, reports the availability of 200 thousand courses [8].

These three challenges are prevalent on online platforms hosting lecture videos: Metadata-based lecture video search, searching within lecture videos, and answering questions.

First, regardless of the platform, universities, companies, and other users who upload videos are usually required to provide metadata, e.g., a title, a description, a category, or other information, which should allow users to find the uploaded contents on the platform [20]. For example, a lecture video on Natural Language Processing (NLP) might appear in a category named Machine Learning (ML), or when said keyword is entered into a search field. Given the ever-increasing supply of online learning videos, learners automatically rely more heavily on such manually entered metadata, which could be incomplete, during their search process.

Second, a closely related problem is that when one is attempting to answer a specific question, even after finding a relevant video, an answer could lie anywhere in the video. Subsequently, whether the right video was found based on the search on metadata might only become clear once the entire video has been watched.

Third, assuming popular lecture video platforms allowed the search of videos and relevant sequences based on their contents, users would still need to watch sequences from potentially various videos, and then combine and understand contextual information in order to formulate an answer. Video Question Answering (VQA) is a topic that has been extensively researched in the recent past [18]. Most related research focuses on narrative, non-educational video data, which leads to research focusing on Computer Vision (CV) techniques [18].

The objective of this research is to investigate the potential of Large Language Models (LLMs) and relevant auxiliary techniques, particularly Retrieval-Augmented Generation (RAG), in addressing challenges specific to lecture video question answering. The resulting framework was developed as an Artificial Intelligence (AI) system that integrates LLMs, RAG with advanced techniques, and conventional techniques to extract text from audio and image content of lecture videos.

The research offers the following contributions: (1) a system architecture for a fully locally run lecture video question answering system using RAG, (2) an assessment of the overall effectiveness of the resulting RAG system which utilizes textual context extracted from lecture videos to improve answer correctness, (3) research findings and an assessment of evaluation techniques for lecture video RAG systems based on their usefulness in this context, (4) a comparison of techniques for integrating multimodal embeddings derived from image and audio content in lecture videos: retrieving the most relevant context from both modalities independently, forcing retrieval from both modalities for each most relevant retrieved context element, and averaging embeddings prior to retrieval, and (5) a comparison of three retrieval configurations in lecture video RAG systems: retrieval, retrieval with re-ranking, and retrieval with context filtering using an LLM.

In general, this research provides practical insights into the design and optimization of a RAG system in the lecture video domain.

2 Related Work

The search problems described in the introduction have been addressed by several researchers. Repp et al. [30] extract text from video lectures as well as lecture slides, and videos are manually segmented. A knowledge base is created containing text that has been processed using traditional NLP techniques such as removing stop-words and stemming.

Sercan Ağzıyağlı and Oğul [31] use OCR to extract text from videos and a Bi-LSTM model to classify these videos. Chand and Oğul [4] utilize ASR to extract text from video based on its audio. In [31], the focus is on classifying lecture videos using text extracted via OCR. In [4], the extracted text from ASR is used for segmenting videos and identifying segment boundaries.

Yang and Meinel [37] and Sreepathy [34] use ASR and OCR to extract text. In [37], videos are automatically segmented, text is extract from image and audio, and a system is built which allows searching for keywords in videos and display relevant videos and sequences related to the search term. Similarly, [34] uses ASR and OCR to extract text and search videos by keyword - however, it is not specified whether a corpus of videos can be searched at once or if search is possible per video only.

Khurana and Deshpande provide a comprehensive overview of VQA techniques and illustrate a trend from text-based methods to content-based methods because text-based methods do not provide context (actions, relationships between events) [18]. Content-based methods rely on CV techniques to capture that context. The same article also mentions that text-based methods are more suitable for automatic generation of subtitles, news videos, and educational videos.

Several methods have been proposed for VQA. Colas et al. propose a new dataset called TutorialVQA [9]: Tutorial videos are manually segmented into 6000 video question answer examples. An LSTM model is trained to learn to answer the questions. Ko et al. propose a model that achieved state-of-the-art results on several VQA datasets [19]: LLaMA is fine-tuned by training with video and question pairs to predict an answer, VQ $\to$ A, and in flipped settings VA $\to$ Q and QA $\to$ V.

Finally, Lewis et al. introduce the concept of retrieval-augmented generation (RAG), which proposes a model architecture where a retriever module, able to add information from outside the model's parameters, is placed in a sequence to sequence model to improve generated outputs [21]. More advanced RAG techniques have been proposed, such as Self-RAG, where a language model is trained to retrieve context only when required, and to determine its output's utility [3], as well as Corrective RAG, which introduces a retrieval evaluator, based on which context can either be refined and used, replaced with information from a web search, or a combination of the two strategies [36].

3 Methods

We developed a lecture video RAG system based on the research on text extraction from lecture videos, as well as research regarding RAG techniques. We chose ASR as the technique to transcribe spoken word from the lecture videos' audio data, and OCR to extract text from the lecture videos' image data.

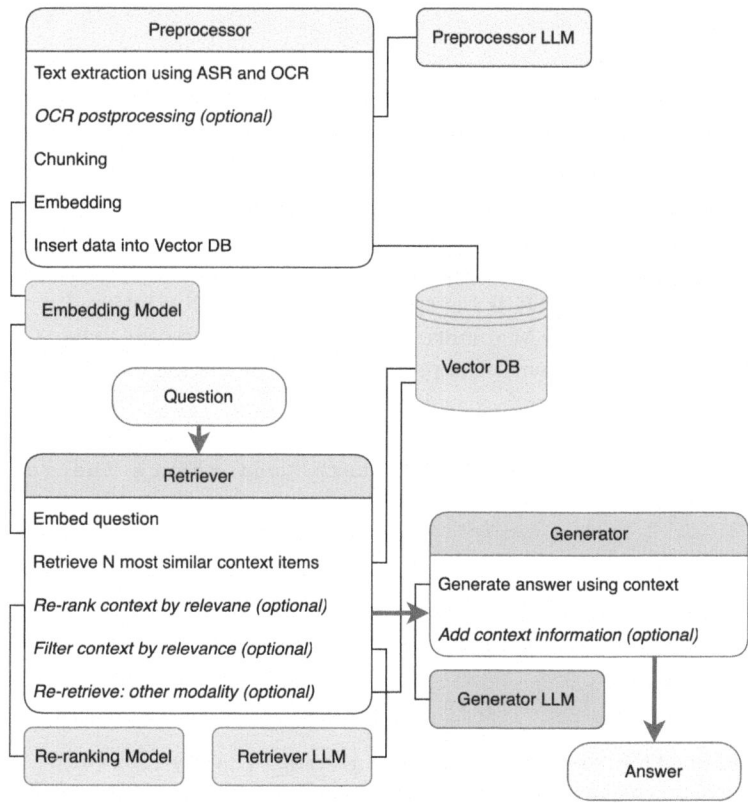

Fig. 1. System Overview

3.1 Development Environment

The project was implemented using Python, and freely available libraries and frameworks were used to facilitate the development of the system. A custom environment was set up using the PDM package manager, ensuring consistent dependency management [29]. For GPU-accelerated LLM use, Llama-CPP-Python is used, supporting both macOS M-family GPUs and CUDA GPUs [13].

3.2 System Overview

Figure 1 shows the system and the most important logical steps involved in preprocessing data to add lecture videos to the vector database, how the question is used to retrieve relevant context, and finally how the system generates an answer.

3.3 Preprocessing

Text Extraction from Speech — We chose OpenAI's Whisper library, after a brief research of existing ASR libraries, due to its ease of use, various model sizes and its performance [28].

Text Extraction from Image — An image of the lecture videos is captured every n seconds, and OCR is performed on the captured image if the grayscale pixel difference to the previous capture is above a threshold. We chose 'Tesseract' and JaidedAI's 'EasyOCR' as OCR libraries [17,32]. Optionally, 'Llama 3 8B Instruct Nous' (Q5 K M quantized) is used for post-processing OCR text (hyperparameter setting) with the prompt shown in Listing 1.1.

```
"""Act as a post-processing agent for OCR text. Your task is
    to clean and improve the readability of the given OCR
    text. Use the provided 'OCR text' and return the refined
    version.

*Example*

Input: 'OCR text:
This iss an exsmple of OCR tex.'

Output: 'This is an example of OCR text.'

Ensure that the cleaned text is grammatically correct,
    properly formatted, and free of OCR errors or artifacts.
    Maintain the original meaning and context of the text."""
```

Listing 1.1. OCR postprocessing prompt

Chunking of Timestamped Text — Extracted text is split into sentences while preserving timestamps (start and end of text). Sentences are then combined and overlapped. Hyperparameters define the maximum character length of combined sentences and maximum character length of the overlap.

Text Embedding — Text is embedded using the RAG system's embedding model. A hyperparameter controls whether embeddings of text from ASR and OCR are embedded and inserted into the vector database independently, or if their vectors are averaged before insertion.

Storing Data — The text data, embeddings of text, as well as metadata (timestamps, text source modality [ASR or OCR], course name, video name) are stored in the RAG system's vector database.

Additional, notable preprocessing hyperparameters are the Whisper model, and OCR hyperparameters (capture every n seconds, difference threshold).

3.4 Retrieval

Embedding — The input question is embedded using an embedding model. Informed by the Massive Text Embedding Benchmark (MTEB) [24], we chose Nomic AI's 'nomic-embed-text-v1', which fits our criteria of an embedding model under a permissive open source license with at least 4000 tokens context length, below 1B parameters [27].

Vector Store — Using the vector store, a nearest neighbor search is performed based on the cosine distance to retrieve n most similar context items (text chunks), specified by a hyperparameter.

Post-retrieval — A hyperparameter allows re-ranking using a post-retrieval model which compares each retrieved document with the question and orders by descending similarity to keep the n most relevant documents, and the ability to define how many documents are retrieved before re-ranking [15,26]. Another hyperparameter allows filtering relevant context using a 'Llama 3 8B Instruct Nous' (Q5 K M quantized) in the retriever module. Finally, the retriever also allows re-retrieving context from the other modality for each retrieved document (e.g., if the first retrieved document is from ASR, find all OCR documents within the same timeframe).

3.5 Generation

Retrieved context is prepended to the question, based on which an answer is generated using a pre-trained LLM. Metadata of the retrieved context is summarized and added to the generated answer. We chose 'Mistral 7B Instruct v0.2' (Q5 K M quantized) as generation model. Listing 1.2 shows the generation prompt.

```
f"""Given the context, answer the question.

*Context:*
{context}

*Question:*
{question}
"""
```

Listing 1.2. Generation Prompt

3.6 Individual Component Evaluation

Certain components of the system can be evaluated and tuned individually. ASR can be evaluated using the Word Error Rate (WER), which measures the percentage of words incorrectly recognized: the sum of substitutions, insertions, and deletions divided by the number of words in the reference text [2].

$$\text{WER} = \frac{\text{Substitutions} + \text{Deletions} + \text{Insertions}}{\text{Total number of words}} \tag{1}$$

Audio transcripts are available for each video, and serve as ground truth for ASR WER calculation. Due to WER's simplicity, it was used to find initial OCR hyperparameters with the WER metric as well by using lecture slides, available for most videos as PDFs with selectable text, which allows the transcription to text files which serve as ground truth to compute OCR WER.

The ability to evaluate ASR and OCR components individually enables a pre-tuning of the system and eliminating a part of the hyperparameter optimization search space.

3.7 RAG Evaluation

As no universally recognized standard for RAG evaluation exists, the reference-free metrics from the Retrieval-Augmented Generation Assessment (RAGAs) framework, which are mentioned frequently according to our brief online research [11], were explored and an attempt to tune the system using these metrics was made. RAGAs introduces the retrieval metric Context Relevance, which assesses how well the retrieved context relates to the question, and the generation metrics Answer Relevance, which evaluates how well the answer covers the question and Faithfulness, which measures how strongly the answer is based on the provided context [11]. The corresponding Ragas library also features the reference-based Answer Correctness metric, which allows for end-to-end evaluation [12]. Additionally, another open source library, DeepEval, was used to compute the same metrics, in order to increase confidence in the measurements [10].

In a later stage of the research project, however, a custom, reference-based evaluation function using OpenAI's API with their GPT-4o model (version 'gpt-4o-2024-05-13') was implemented to evaluate each generated answer of our system and tune hyperparameters.

3.8 Databases and Caching

The vector database forms a central part of the RAG system. ChromaDB was chosen because of its ease of use and because it is open-source [5]. Chroma allows the local storage of documents and their embeddings, as well as metadata, querying via common distance metrics and metadata-based keyword search [6].

Moreover, in order to cache the outputs of the ASR and OCR components of the RAG system, the open-source database SQLite is hosted locally as part of the preprocessing module [33].

4 Experiments

4.1 Experimental Setup

Experiments were predominantly conducted on high-performance computing (HPC) servers with NVIDIA Quadro RTX 8000 graphic cards. Random seeds were set for reproducibility, and generation temperature 0 was set for LLMs in the RAG system.

4.2 Data

Data from the following lecture videos, which are accompanied by questions and answers in the form of quizzes or exams, were utilized:

- Harvard OpenCourseWare - CS50's Introduction to Artificial Intelligence with Python [14]
- MIT OpenCourseWare 9.00SC - Introduction to Psychology [22]
- MIT OpenCourseWare 8.20 - Introduction To Special Relativity [23]

In order to determine initial hyperparameters for ASR and OCR, before tuning the RAG system, a pre-tuning dataset was manually annotated by randomly selecting three videos per course, downloading their video data, audio transcript and if available, lecture slides. Audio transcripts were manually cleaned of annotations such as the lecturer's name. The Introduction To Special Relativity course videos required some more manual work to obtain a lecture slide: taking screenshots of the video whenever a change occurs, generating a searchable PDF from the screenshots, followed by text extraction.

The main dataset for the RAG system evaluation was created by downloading all videos of the three mentioned courses, totaling 84 videos. Then, questions and answers were taken from the course materials (exams and quizzes). Questions that rely on other questions, tables, graphics, or complex mathematical formulas were ignored. The result is a dataset with 90 question-answer pairs, shown in Table 1. We split the data into an evaluation dataset (80%) and test dataset (20%).

Table 1. Comprehensive Dataset Overview

Course Topic	Videos	Q&A Pairs	Question Types
Artificial Intelligence	8	25	Multiple Choice (4 options): 8 Multiple Choice (5 options): 7 Multiple Choice (6 options): 3 Multiple Choice (7 options): 1 Multiple Choice (8 options): 2 Multiple Choice (11 options): 1 Free Form: 3
Psychology	24	51	Multiple Choice (4 options): 51
Special Relativity	52	14	True/False or Yes/No: 11 Free Form: 3

4.3 Embedding Model

The SBERT embedding model 'allMiniLM-L6-v2' is compared to the chosen Nomic embedding model 'nomic-embed-text-v1', by using a self-made dataset. Furthermore, the usefulness of semantic embeddings to retrieve documents of varying size with short queries is verified with this experiment [16,27].

Algorithm 1 Comparing Nomic and SBERT Embeddings for Wikipedia Article Retrieval

$articles \leftarrow$ downloadRandomWikipediaArticles()
$nomicEmbedder \leftarrow$ initializeNomicEmbedder()
$nomicCollection \leftarrow$ createVectorDBCollection($nomicEmbedder$)
$sbertEmbedder \leftarrow$ initializeSBERTEmbedder()
$sbertCollection \leftarrow$ createVectorDBCollection($sbertEmbedder$)
$llm \leftarrow$ initializeLLM
for $article \in articles$ **do**
    $summary, randomSequence \leftarrow$ generateSummaryAndSequence($article, llm$)
    $articleWithMetadata \leftarrow$ addMetadata($article, summary, randomSequence$)
    embedAndInsert($articleWithMetadata, nomicCollection$)
    embedAndInsert($articleWithMetadata, sbertCollection$)
$results \leftarrow \{\}$
for $collection \in \{nomicCollection, sbertCollection\}$ **do**
    $articleWithMetadata \leftarrow$ getAllArticlesFromCollection(collection)
    for $article \in articles$ **do**
        $summary, randomSequence \leftarrow$ getMetadata($articleWithMetadata$)
        $summaryResults \leftarrow$ retrieveAndEvaluate($summary, collection$)
        $sequenceResults \leftarrow$ retrieveAndEvaluate($randomSequence, collection$)
        $results \leftarrow$ appendResults($results, summaryResults, sequenceResults$)

Our evaluation depicted in Algorithm 1 for the embedding models downloads a specified number of randomly chosen Wikipedia articles as markdown-formatted text. These articles are embedded and stored in a vector store, in a different collection per embedding model. As a query, a summary of max. 150 characters is created of a random sequence of 300 characters, generated by the Mistral model. For each article, we also extract a random sequence of 100 characters as a second query. We then embed each query, retrieve the 3 nearest neighbors, and compute recall@1 up to recall@3. The random sequence is 100 characters long. The summary is max. 150 characters long and summarizes another random sequence of 300 characters of the original text.

4.4 Isolated Pre-tuning: ASR, OCR

We evaluate Whisper models by computing the WER per model using the transcripts as ground truth. A fixed range of OCR capture intervals and difference thresholds are compared by computing the WER using the lecture slides as ground truth. Therefore, a script was developed which contains paths to each video file, transcript, and lecture slide document for each item in the ASR and OCR pre-tuning dataset.

Extensively tuning OCR hyperparameters with this procedure would misfit the configuration because slides containing no repetition serve as ground truth, while captured images most likely contain repetitive sequences; a lecturer could switch back and forth between slides, walk in front of the slides, the display of slides may be interrupted during the course of the lesson, or transition effects might result in contents appearing sequentially rather than all at once. The potentially discontinuous or repetitive nature of lecture slide contents limits the usefulness of this tuning step for OCR. However, the pre-tuning procedure translates well to the RAG system for ASR due to the continuous nature of extracted text from audio, containing no repetitions. For ASR and OCR, chunk size and overlap are additional hyperparameters that are tuned later in the RAG tuning step.

The ASR and OCR preliminary results and an exploratory online investigation on chunk sizes and overlaps lead to the establishment of an informed default hyperparameter configuration: Whisper model medium.en, 5000 character chunk size and 500 character overlap for ASR, OCR capturing every 180 s with a min. 30% frame difference threshold, 1000 character chunk size, 100 character overlap, retrieve 5 context documents with a max. cosine distance of 1.

4.5 RAG System Evaluation and Tuning

We initially used the 'Ragas' library to evaluate the correctness of the system's outputs, complemented by another open-source library, 'DeepEval', to increase confidence in our measurements. These evaluations relied on strong LLMs capable of returning JSON-formatted outputs, specifically 'Llama 3 8B Instruct' (Q5 K M quantized) for the Ragas library and 'Mixtral 46B Nous Hermes 2 DPO' (Q5 K M quantized) for the DeepEval library.

However, the average answer correctness metric from these libraries, which ranged from 0.0 to 1.0, proved suboptimal to tune our system. We theorize this is because incorrect answers containing some truthful statements would still receive a score above 0.0, leading to less effective tuning.

To address this, we developed our own evaluation function using OpenAI's API and their GPT-4o model and the prompt shown in Listing 1.3, enforcing a binary classification (1 for correct and 0 for incorrect) of the answers. This approach was more efficient and precise compared to running resource-intensive evaluation models locally, which slowed down the tuning process. We employed Optuna for hyperparameter optimization [1].

```
f"""
Question: {question}
Ground Truth (Correct) Answer: {ground_truth}
Generated Answer: {generated_answer}
```

```
 6  Evaluate the generated answer in comparison to the ground
        truth.
 7  Respond with 1 if the generated answer is correct,
 8  otherwise respond with 0.
 9  You MUST return 1 or 0, and nothing else.
10  It is VERY important that your output is only 1 or 0.
11
12  Example Response A:
13  1
14
15  Example Response B:
16  0
17  """
```

Listing 1.3. Evaluation prompt used with OpenAI's GPT-4o model

Lastly, the retrieval accuracy of the RAG system is evaluated by computing the percentage of retrieved context elements which originated from the same course as the question for which they were retrieved.

5 Results

5.1 Embedding Model

The Nomic model scored better at creating meaningful embeddings for random sequences and summaries of Wikipedia articles: Using the vector database's cosine nearest neighbor search, more of the retrieved articles are the correct source for the summary and random sequence than with the default SBERT embedding model. A total of 7237 articles were used. Results in Fig. 2 show that Nomic summaries and especially random sequences outperform the SBERT model.

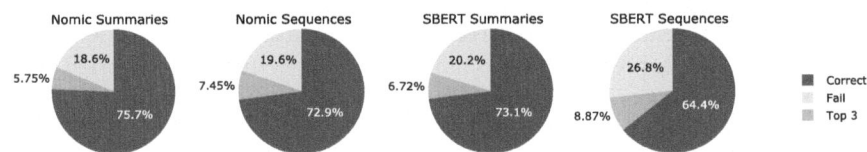

Fig. 2. Embedding Models - Retrieval Comparison

5.2 Isolated Pre-Tuning: ASR, OCR

The Whisper models tiny.en and medium.en stood out; the WER of the smallest model tiny.en is comparable to that of the largest models, and medium.en because it resulted in the lowest WER, as shown in Fig. 3.

For OCR, capture every 180 s with 30% frame difference threshold performed best. The explored hyperparameter space is shown in Fig. 4.

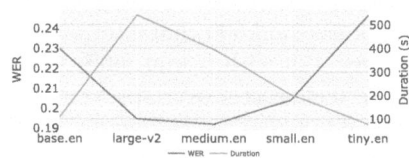 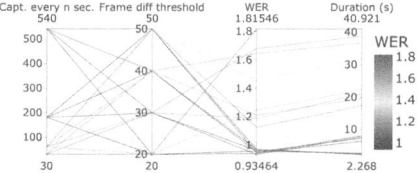

Fig. 3. ASR - WER and duration in seconds per Whisper model

Fig. 4. OCR - WER per configuration

5.3 RAG System Evaluation and Tuning

Our RAG system achieves an average answer correctness of 38.89% on the test dataset without context (cosine distance threshold set to 0). After tuning, this increases to 54.44%, representing a 40% improvement. For comparison, OpenAI's GPT-3.5 Turbo (version 'gpt-3.5-turbo-0125') achieves 50.00% correctness on the same test dataset. We observe comparable performance also on the larger evaluation dataset, where our system achieves 50.83% accuracy and GPT-3.5 Turbo reaches 51.11%. Therefore, it can be concluded that *a RAG system based on textual context extracted from lecture videos is effective at improving the average accuracy of generated answers.*

Figure 5 shows how many percent of the top-k retrieved documents stem from the same course as the question (course, in this context, refers to one of three courses). The top-1 course retrieval accuracy is 83.33% and drops to 76.39% until top-8 with the hyperparameter configuration with the highest answer correctness.

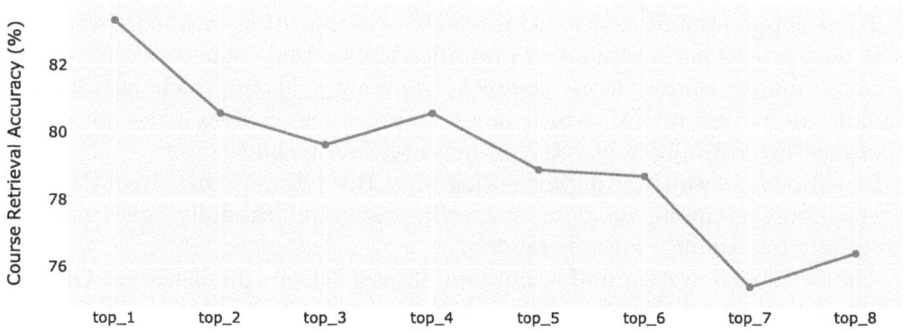

Fig. 5. Course Retrieval Accuracy per Top-N

Notably, the accuracy decreases with the amount of context elements. The correctness of the specific video source for retrieved context was not measured due to a lack of ground truth regarding this. We theorize that a high course retrieval accuracy does not necessarily indicate a better RAG system, because knowledge regarding a particular question could be contained in any course.

These result were achieved with the following hyperparameter configuration: Whisper model medium.en, 5000 character chunk size and 500 character overlap for ASR, OCR library EasyOCR capturing every 45 s with a min. 60% frame difference threshold, 2500 character chunk size, 750 character overlap, retrieving 40 context elements with a max. cosine distance of 0.8, and re-ranking 8 context elements using the mixedbread ai's mxbai-rerank-large-v1 model.

The evaluation of the end-to-end system further suggests:

- In the context of our dataset, boolean answer evaluation is more useful for hyperparameter tuning of a RAG system than the assessment of the overall correctness of generated answers.
- Among the techniques for integrating multimodal embeddings, independently retrieving the most relevant context across modalities yields the highest answer correctness.
- Retrieval followed by re-ranking using a re-ranking model performed best among the explored retrieval methods

6 Conclusions

This study introduces a multimodal framework to apply advanced RAG techniques to the domain of lecture videos. The results are promising and suggest the usefulness of RAG in domains that are not text-native. However, the following limitations may restrict the generalizability of the study's results: the dataset comprises only three courses and topics, which may not represent a broader educational spectrum; the evaluation robustness during hyperparameter tuning is limited by single-run answer accuracy evaluations by a proprietary LLM; and the evaluation of the system's performance could be extended by incorporating more reference-based metrics.

There is potential for continued research: experimenting with more advanced OCR post-processing techniques to improve the textual context quality of this modality; implementing more complex chunking schemes, such as semantic chunking described in [25], which might improve the context's relevance; and experimenting with different retrieval methods and technologies.

In a practical setting, implementing this RAG framework should involve hyperparameter tuning on domain-specific data and carefully selecting, and potentially fine-tuning, suitable models.

The developed system and annotated dataset will be published on GitHub: github.com/repetitioestmaterstudiorum/lecture-video-rag.

Disclosure of Interests. The authors have no competing interests to declare that are relevant to the content of this article.

References

1. Akiba, T., Sano, S., Yanase, T., Ohta, T., Koyama, M.: Optuna: a next-generation hyperparameter optimization framework. In: The 25th ACM SIGKDD International Conference on Knowledge Discovery & Data Mining, pp. 2623–2631 (2019)

2. Ali, A., Renals, S.: Word error rate estimation for speech recognition: e-WER. In: Proceedings of the 56th Annual Meeting of the Association for Computational Linguistics (Volume 2: Short Papers), pp. 20–24. Association for Computational Linguistics, Melbourne, Australia (2018). https://doi.org/10.18653/v1/P18-2004. http://aclweb.org/anthology/P18-2004
3. Asai, A., Wu, Z., Wang, Y., Sil, A., Hajishirzi, H.: Self-RAG: Learning to Retrieve, Generate, and Critique through Self-Reflection (2023). https://doi.org/10.48550/arXiv.2310.11511
4. Chand, D., Oğul, H.: A framework for lecture video segmentation from extracted speech content. In: 2021 IEEE 19th World Symposium on Applied Machine Intelligence and Informatics (SAMI), pp. 000299–000304 (2021). https://doi.org/10.1109/SAMI50585.2021.9378632. https://ieeexplore.ieee.org/document/9378632
5. Chroma: chroma-core/chroma (2024). https://github.com/chroma-core/chroma
6. Chroma: Python Collection | Chroma Docs (2024). https://docs.trychroma.com/reference/py-collection
7. Class Central: By The Numbers: MOOCs in 2021 (2021). https://www.classcentral.com/report/mooc-stats-2021/
8. Class Central: About (2023). https://www.classcentral.com/about
9. Colas, A., Kim, S., Dernoncourt, F., Gupte, S., Wang, D.Z., Kim, D.S.: TutorialVQA: question answering dataset for tutorial videos (2020). https://doi.org/10.48550/arXiv.1912.01046, version: 2
10. Confident AI: confident-ai/deepeval (2024). https://github.com/confident-ai/deepeval
11. Es, S., James, J., Espinosa-Anke, L., Schockaert, S.: RAGAS: Automated Evaluation of Retrieval Augmented Generation (2023). https://doi.org/10.48550/arXiv.2309.15217
12. Exploding Gradients: Ragas Answer Correctness Metric (2024). https://docs.ragas.io/en/latest/concepts/metrics/answer_correctness.html
13. Gerganov, G.: ggerganov/llama.cpp (2024). https://github.com/ggerganov/llama.cpp
14. Harvard: CS50's Introduction to Artificial Intelligence with Python (2024). https://cs50.harvard.edu/ai/2023/
15. HuggingFace: mixedbread-ai/mxbai-rerank-large-v1 · Hugging Face (2024). https://huggingface.co/mixedbread-ai/mxbai-rerank-large-v1
16. HuggingFace: sentence-transformers/all-MiniLM-L6-v2 · Hugging Face (2024). https://huggingface.co/sentence-transformers/all-MiniLM-L6-v2
17. JaidedAI: EasyOCR (2020). https://github.com/JaidedAI/EasyOCR
18. Khurana, K., Deshpande, U.: Video question-answering techniques, benchmark datasets and evaluation metrics leveraging video captioning: a comprehensive survey. IEEE Access **9**, 43799–43823 (2021). https://doi.org/10.1109/ACCESS.2021.3058248. https://ieeexplore.ieee.org/document/9350580
19. Ko, D., Lee, J.S., Kang, W., Roh, B., Kim, H.J.: Large Language Models are Temporal and Causal Reasoners for Video Question Answering (2023). https://doi.org/10.48550/arXiv.2310.15747
20. Kumbham, S., Debnath, A., Rao, K.S.: Efficient Indexing of Meta-Data (Extracted from Educational Videos) (2023). https://doi.org/10.48550/arXiv.2401.01356, [cs]
21. Lewis, P., et al.: Retrieval-Augmented Generation for Knowledge-Intensive NLP Tasks (2021). https://doi.org/10.48550/arXiv.2005.11401, version: 4
22. MIT: Introduction to Psychology | Brain and Cognitive Sciences (2024). https://ocw.mit.edu/courses/9-00sc-introduction-to-psychology-fall-2011/

23. MIT: Introduction to Special Relativity | Physics (2024). https://ocw.mit.edu/courses/8-20-introduction-to-special-relativity-january-iap-2021/
24. Muennighoff, N., Tazi, N., Magne, L., Reimers, N.: MTEB: Massive Text Embedding Benchmark (2023). https://doi.org/10.48550/arXiv.2210.07316
25. Muszynska, E.: Semantic chunking. University of Cambridge Repository (2020). https://doi.org/10.17863/CAM.59299. https://www.repository.cam.ac.uk/handle/1810/312207
26. Nogueira, R., Cho, K.: Passage Re-ranking with BERT (2020). https://doi.org/10.48550/arXiv.1901.04085
27. Nussbaum, Z., Morris, J.X., Duderstadt, B., Mulyar, A.: Nomic Embed: Training a Reproducible Long Context Text Embedder (2024). https://doi.org/10.48550/arXiv.2402.01613
28. OpenAI: openai/whisper (2024). https://github.com/openai/whisper
29. PDM: pdm-project/pdm (2024). https://github.com/pdm-project/pdm
30. Repp, S., Linckels, S., Meinel, C.: Question answering from lecture videos based on an automatic semantic annotation. In: Proceedings of the 13th Annual Conference on Innovation and Technology in Computer Science Education, ITiCSE 2008, pp. 17–21. Association for Computing Machinery, New York (2008). https://doi.org/10.1145/1384271.1384278
31. Sercan Ağzıyağlı, V., Oğul, H.: Multi-level lecture video classification using text content. In: 2020 IEEE 14th International Conference on Application of Information and Communication Technologies (AICT), pp. 1–5 (2020). https://doi.org/10.1109/AICT50176.2020.9368692. https://ieeexplore.ieee.org/document/9368692. ISSN 2472-8586
32. Smith, R.: An overview of the tesseract OCR engine. In: Ninth International Conference on Document Analysis and Recognition (ICDAR 2007), vol. 2, pp. 629–633 (2007). https://doi.org/10.1109/ICDAR.2007.4376991. https://ieeexplore.ieee.org/document/4376991. ISSN 2379-2140
33. SQLite: sqlite/sqlite (2024). https://github.com/sqlite/sqlite
34. Sreepathy, G.: Automated analysis and indexing of lecture videos. In: Iowa State University (2020). https://doi.org/10.31274/etd-20210114-142. https://lib.dr.iastate.edu/etd/18407
35. Wikipedia: Massive open online course (2023). https://en.wikipedia.org/w/index.php?title=Massive_open_online_course&oldid=1171690342, page Version ID: 1171690342
36. Yan, S.Q., Gu, J.C., Zhu, Y., Ling, Z.H.: Corrective Retrieval Augmented Generation (2024). https://doi.org/10.48550/arXiv.2401.15884, version: 1
37. Yang, H., Meinel, C.: Content based lecture video retrieval using speech and video text information. IEEE Trans. Learn. Technol. **7**(2), 142–154 (2014). https://doi.org/10.1109/TLT.2014.2307305. https://ieeexplore.ieee.org/document/6750040

Agent-Based Simulation Leveraging Declarative Modeling for Efficient Resource Allocation in Emergency Scenarios

Ionuţ Murareţu[✉], Alexandra Vultureanu-Albişi, Sorin Ilie, and Costin Bădică

Department of Computers and Information Technology, University of Craiova, Craiova, Romania
{ionut.muraretu,alexandra.vultureanu,sorin.ilie,costin.badica}@edu.ucv.ro

Abstract. Emergencies can cause complex challenges for emergency hospitals, requiring an effective allocation of limited resources to reduce the loss toll. In this paper, we address a declarative modeling approach for the problem of assigning casualties to hospitals in an agent-based simulation context for emergency crises. Our approach leverages autonomous and decision-making-capable agents to optimize resource allocation while addressing various constraints, including but not limited to resource shortages and patient prioritization criteria. In addition, we present simulation outcomes that illustrate the effectiveness of the proposed approach in various randomly generated mass casualty incident scenarios. Furthermore, we conclude that our approach offers a promising direction to increase the efficacy and resilience of the rescue team's responses in the face of mass casualty incidents.

Keywords: declarative modeling · optimization · agent-based simulation

1 Introduction

In recent years, Romania has faced numerous challenges regarding its healthcare system, particularly in managing hospital resources, according to [3, 7]. One of Romania's healthcare system's significant challenges is the effective allocation of hospital beds. Mass casualty accidents and emergency crises can easily overwhelm healthcare resources, significantly increasing the demand for hospital beds. This has highlighted the need for better planning and coordination within the healthcare system to ensure that resources are efficiently utilized during times of crisis. When we mention hospital beds, we also refer to the resources needed for a specific treatment.

Many studies have shown the potential benefits of using optimization models in resource allocation during times of crisis. These models are often packed as decision support systems that can assist in making critical decisions quickly and efficiently [2,5].

We consider that an agent-based semi-automated system can improve the efficiency of allocating hospital resources so that the survival ability of the injured person can be increased. Many previous research papers demonstrate the feasibility of using agent-based approaches to simulate real-life scenarios to optimize outcomes. In [1], the authors provide the details of developing an intelligent freight broker agent for logistic services. Paper [8] presents an agent-based simulation model for responding to mass casualty incidents (MCI) caused by disasters. An agent-based simulation model was introduced in [6] as a decision-making system to evaluate a mobile-based system designed to support emergency evacuation.

Given the dynamic and unpredictable nature of crisis scenarios, an agent-based modeling approach can be highly beneficial. The two main factors that significantly affect the effectiveness of crisis management are the temporal dimension and treatment capabilities. We postulate that implementing a more advanced system could also help reduce emergency response times. Although optimization models can benefit resource allocation, solely depending on them may neglect the crucial role of human judgment and adaptability in crisis scenarios, presenting routes for further research. A semi-automated agent-based system could offer a more balanced approach by incorporating optimization models and human judgment.

Incorporating declarative modeling into agent-based systems to handle emergency resource allocation provides the ability to manage the complexity of high-level constraints and rules, reducing the load on decision-makers and ensuring that the system can perform in various scenarios efficiently. The potential of declarative modeling for solving complex problems efficiently has been demonstrated in previous research studies [4].

In this paper, we consider the feasibility of integrating an optimization model in an agent-based simulation context to improve the use of hospital beds as resources during crises and test the system further in real-life scenarios. Simulation models can address the dynamic and unpredictable nature of crisis scenarios. Traditional optimization models may miss the crucial role of human adaptability in emergencies, whereas a semi-automated agent-based system balances computational optimization with human decision. Furthermore, the agent-based simulation allows testing and refining resource allocation strategies before implementation in real-world scenarios. This method allows for generating and evaluating various crisis scenarios, supporting the identification of potential weaknesses and areas of improvement in resource allocation. The paper introduces an agent-based simulation context and provides a mixed-integer linear programming formulation for the casualty-hospital assignment problem. We conducted the evaluations by considering emergency hospitals in each county in Romania with randomly generated resources, aiming to assess different emergency scenarios.

The paper is structured as follows: Sect. 2 depicts an agent-based system architecture and defines all the involved agents. Furthermore, we present the mixed-integer linear programming model of the initial casualty-hospital assignment. Section 3 provides an initial experimental setup with some preliminary results. Section 4 concludes the paper.

2 Problem Formulation

Given the hospitals and their available resources in a specific geographical region, we consider the problem of handling emergency scenarios within the region. The objective is to minimize the number of deaths for each casualty of an incident by using the available resources to manage the casualty in time to increase the chances of survival.

We propose a centralized agent-based simulation model that consists of three main entities that interact: ambulances, hospitals and the Incident Commander (IC). The IC is the system's central unit that manages the hospital resources and schedules ambulances.

In the proposed model, the ambulances are critical in rapidly transporting casualties to the appropriate medical facilities. The IC, equipped with real-time information on the location and status of each ambulance, dispatches them to incident sites and hospitals based on factors such as proximity, bed capacity, and available treatment type capabilities. This dynamic routing ensures that no hospital is overwhelmed while optimizing travel times, reducing delays in treatment, and preventing resource bottlenecks. By integrating information about each casualty's condition, required treatment type, and hospital availability, the IC can prioritize severe cases and ensure that resources are allocated efficiently to decrease lost casualties. This centralized approach allows for coordinated decision-making in high-pressure, resource-constrained environments, significantly enhancing the emergency response process.

For a better understanding of where the system comes in action, we first detail the events of a disaster in chronological order:

1. The event occurs, and teams are dispatched to the scene.
2. Reports come in via phone calls from civilians. The phone operator performs triage and sends a casualty report to the IC.
3. The rescue teams immediately start to report casualties to the IC.
4. Here is the point where the proposed system comes into action. IC schedules the available ambulances to handle casualties and transport them to the designated hospital. The designated hospital is selected based on the availability of beds to treat the casualty injury type.

2.1 System Design

Incident Commander. The IC entity is characterized by a list of hospitals and casualties in the form of raw data. Its capabilities include calculating route

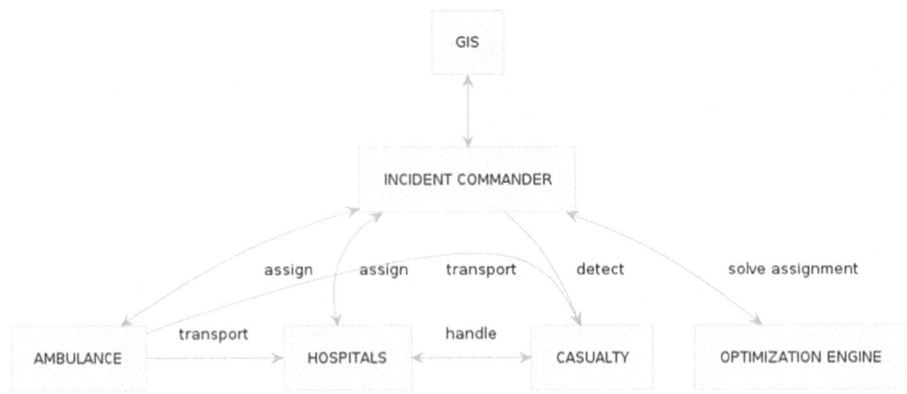

Fig. 1. Agent-based model: agents and interactions

time based on information from a Geographical Information System (GIS) and generating schedules for ambulances to attend to casualties. This central entity allows ambulances to cooperate if the scenario requires it (Fig. 1).

Hospitals. A hospital (H) is characterized by a location, a list of available ambulances and beds equipped for different treatment types.

In the presented simulation model, the hospital represents a single agent in an agent population. Each hospital has an ambulance fleet. When IC assigns a hospital to handle a casualty. If the hospital has an available ambulance, it sends it to the casualty; otherwise, the IC tries to get an available ambulance from the closest hospital.

The list of available beds is continuously refreshed as beds are occupied and vacated.

Ambulance. The Ambulance (A) is the only moving agent in the proposed system. It is characterized by its availability status, home hospital location and the ability to transport a single patient simultaneously.

Casualty. The Casualty agent represents a casualty that needs to be addressed when an incident event is triggered. The following attributes define it:

- *CasualtyID*, numerical $1 \ldots n$ where n is the total number of casualties
- *Location*, lat and long to find transport time to nearest suitably equipped bed
- *Treatment Type* identifies the bed equipment required by the type of injury.
- *ProcessingTime[min]* is the estimated time for processing a casualty and includes rescue time, first-aid time and transportation time

Optimization Engine. The optimization engine is responsible for solving the hospital-casualty assignment problem. The solution provided by the module is used to dispatch the closest available ambulance to transport the casualty to the designated hospital.

Interaction Protocol. As depicted in Fig. 2 the possible interactions of the entities are:

- the IC receives reports from the disaster area outside the model
- the IC requests currently available bed list from all hospitals
- the IC notifies each ambulance to perform one of its capabilities on a casualty

Figure 2 represents the sequence diagram of the protocol for a centralized scheduler triggered for each emergency event report.

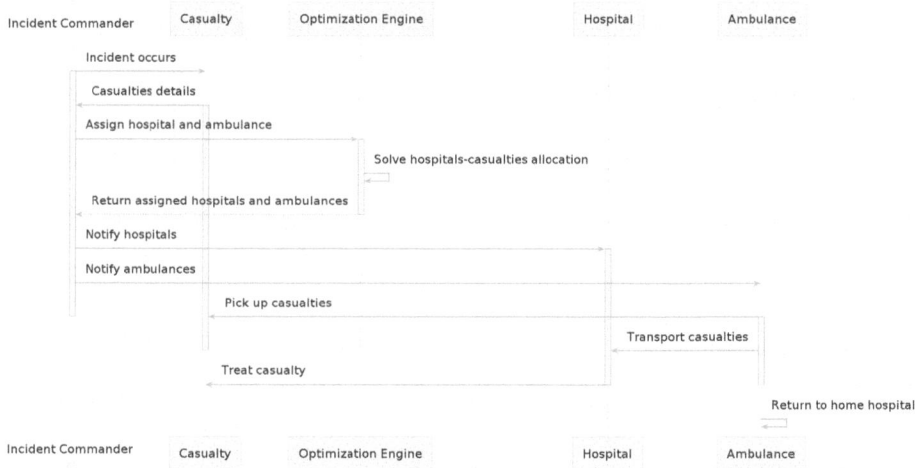

Fig. 2. Interaction protocol - sequence diagram

To test and evaluate various emergency scenarios, we generated data similar to real-life data, such as the geographical location of healthcare facilities, the number of available ambulances and the available beds for each type of trauma.

2.2 Hospital-Casualty Assignment Problem

Let us consider the matrix $\omega \in |C| \times |H|$:

$$\omega = \begin{bmatrix} \omega_{11} & \omega_{12} & \cdots & \omega_{1h} \\ \vdots & \vdots & \ddots & \vdots \\ \omega_{c1} & \omega_{c2} & \cdots & \omega_{ch} \end{bmatrix}$$

where ω_{ch} represents the processing time for casualty c at hospital h. These values are determined based on factors such as travel time, treatment duration, and hospital resources. The objective of the problem is to minimize the number of deaths following an incident. We define the following sets and variables:

Sets:

- H - set of hospitals.
- C - set of casualties.
- τ - set of treatment types.
- B_{ht} - number of available beds for treatment type t at hospital h.
- W_c - maximum waiting time for casualty c.
- T_c - treatment type required by casualty c.

Variables:

- z_c - decision variable indicating whether casualty c dies ($z_c = 1$ if casualty c dies, otherwise $z_c = 0$).
- x_{hct} - decision variable that is 1 if casualty c is assigned to hospital h for treatment type t, and 0 otherwise.

The optimization problem is formulated as follows:

$$\min \sum_{c \in C} z_c \tag{1}$$

$$\text{s.t.} \quad \sum_{c \in C} x_{hct} \leq B_{ht} \quad \forall h \in H, t \in \tau \tag{2}$$

$$x_{hct} = 0 \quad \forall h \in H, \forall c \in C, \forall t \neq T_c \tag{3}$$

$$\sum_{h \in H} x_{hcT_c} \leq 1 \quad \forall c \in C \tag{4}$$

$$\omega_{hc} \cdot x_{hcT_c} \leq W_c + M \cdot z_c \quad \forall h \in H, c \in C \tag{5}$$

$$z_c + \sum_{h \in H} x_{hcT_c} \geq 1 \quad \forall c \in C \tag{6}$$

$$z_c \in \{0, 1\} \quad \forall c \in C \tag{7}$$

$$x_{hct} \in \{0, 1\} \quad \forall h \in H, c \in C, t \in \tau \tag{8}$$

where: Constraint (2) ensures that the number of casualties assigned to hospital h for treatment type t does not exceed the available beds. Constraint (3) assures that a casualty is only assigned to hospitals that provide the required treatment type T_c. Constraint (4) guarantees that each casualty can be assigned to at most one hospital for treatment. Constraint (5) ensures that the processing time of each casualty at hospital h does not exceed the waiting time W_c, unless the casualty dies ($z_c = 1$). Constraint (6) assures that each casualty is either treated by a hospital or marked as deceased. Constraints (7) and (8) define the binary nature of the decision variables.

3 Experiments and Results

3.1 Data

For the sake of simplicity, we consider an emergency hospital in each of Romania's counties based on their real-life locations. Only the resources, not their presence or absence, vary during the simulation.

Table 1 depicts the data related to hospitals. It shows the location of the hospital, geographical coordinates and available beds for the treatment types.

Table 1. Sample of Generated Hospital Data

Hospital	Lat.	Long.	Beds {T1, T2, T3, T4, T5}
Hospital A	45.7321	24.5152	10, 20, 15, 25, 30
Hospital B	46.7543	23.4541	12, 18, 10, 22, 28
Hospital C	44.6432	22.3434	15, 25, 20, 30, 35

Example of Generated Data. Table 2 depicts a sample of generated casualties. It provides the casualty's location as geographical coordinates, the waiting time, and the required treatment type.

Table 2. Sample of Generated Casualty Data

Casualty	Latitude	Longitude	Waiting Time	Treatment Type
C1	45.1234	24.5678	120	T1
C2	45.2345	24.6789	90	T3
C3	45.3456	24.7890	60	T2

Table 3 shows the solution representation of the hospital-casualty.

Table 3. Casualty-Hospital Assignment Solution Representation

Casualty	Hospital	Treatment Type	Lost
C1	Hospital A	T2	No
C2	None	T2	Yes
C3	Hospital B	T5	No

3.2 Scenarios

We defined several scenarios to observe how the proposed model functions. These scenarios were created to simulate different resource allocations for casualties in various situations. Our experiments involve five treatment types, each with a range of waiting times. Every casualty generated is assigned a random treatment type and a random waiting time within that type's designated range. When creating the cost matrix, we consider travel times between casualties and all hospitals. We calculate travel time based on the distance ambulances travel at 80 km/h, which means they must cover to reach hospitals and casualties. Additionally, we introduced a road factor to simulate travel conditions, providing a more realistic estimate of travel times since ambulances do not travel in straight lines.

In the first scenario, the casualties were *uniformly distributed* across an entire geographical area. This scenario aimed to comprehensively understand the model's performance under uniform casualty distribution. Furthermore, this scenario can contribute to understanding how well the hospital network is prepared to handle spreading emergencies by testing the efficiency of bed allocation across a broad area. Additionally, this scenario can offer valuable insights into average response time and resource usage.

The second scenario describes *clustered casualties* in high-density urban areas. This scenario mirrors a realistic urban emergency scenario, where hospitals may face overcrowding due to many casualties within a limited geographic area. This scenario aims to evaluate the capacity of the hospitals in urban centers to absorb a sudden influx of patients.

A third scenario is defined by a *sparse casualty pattern* that simulates emergencies in remote regions, providing valuable insights into the impact of travel times on patient outcomes.

The fourth scenario presents *casualties concentrated in multiple clusters* across a geographical area, simulating multiple incidents happening simultaneously in different locations.

The fifth scenario distributes casualties uniformly but with *limited hospital beds*. This scenario simulates a situation where the hospitals are already strained with other emergencies. The goal of this scenario is to provide ways of developing strategies for patient prioritization and resource allocation during an emergency crisis.

A visual representation of the casualties and hospital distributions can be observed in Figure 3.

3.3 Metrics

To assess the performance of the optimization model across various scenarios, the following metrics have been defined:

The following metrics have been defined to evaluate the optimization model across all the defined scenarios:

- **Average Transportation Time per Casualty:** The average transportation time across all casualties.

$$T_{avg} = \frac{1}{C} \sum_{i=1}^{C} t_i \qquad (9)$$

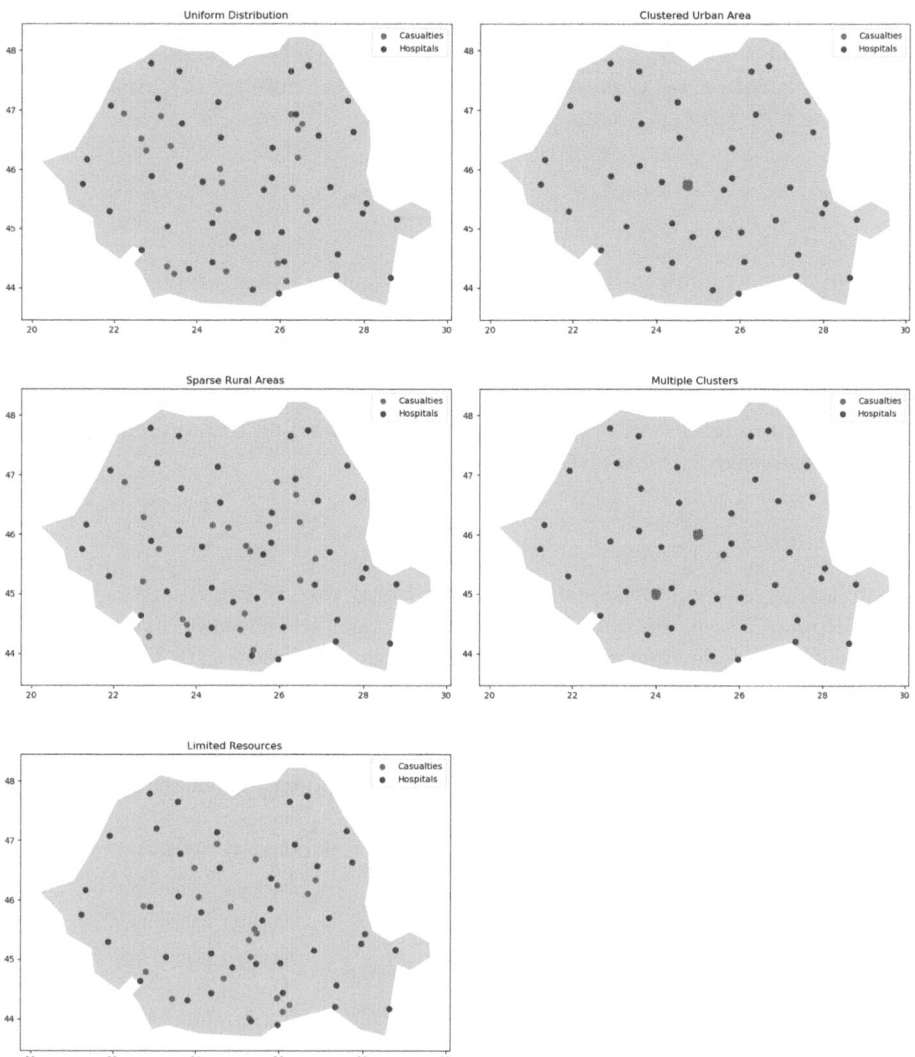

Fig. 3. Example of distribution of hospitals and casualties for each scenario

where C is the total number of casualties and t_i is the transportation time for casualty i.

- **Average Bed Utilization Rate per Hospital:** The average utilization rate of beds per hospital.

$$B_{avg} = \frac{1}{H} \sum_{j=1}^{H} \left(\frac{B_{\text{used},j}}{B_{\text{total},j}} \right) \tag{10}$$

where H is the total number of hospitals, $B_{\text{used},j}$ is the number of beds used at hospital j, and $B_{\text{total},j}$ is the total number of beds available at hospital j.
- **Total Number of Casualties Lost:** The total number of casualties who could not be treated within their waiting time.

$$C_{lost} = \sum_{i=1}^{N} d_i \tag{11}$$

where d_i is a binary variable that is 1 if casualty i is lost and 0 otherwise.

These metrics provide a valuable assessment of the optimization model's performance, focusing on critical aspects such as response time, resource utilization, and casualty outcomes. By analyzing these metrics, we can identify the model's strengths and areas for improvement, leading to more effective and efficient emergency response strategies.

3.4 Results

The following section presents the experimental results for emergency events inside Romania's geographical area with 50 casualties within five treatment types, with the following processing time ranges in minutes {10–30, 30–60, 60–120, 120–180, 180–240}. To properly depict the transportation times, we set the transportation speed of ambulances to 80 km/h. Since we used Haversine distance, a load factor of 1.3 was introduced to approximate the actual travel distance and time between two points in the environment, considering that roads usually do not follow a straight line.

The clustered scenario includes information on generated casualties with the following attributes:

- **Waiting Time Range:** 37 to 478 min
- **Treatment Type Distribution:** T1: 9, T2: 15, T3: 13, T4: 7, T5: 6

For the sparsely distributed casualties scenario, the following were defined:

- **Waiting Time Range:** 40 to 460 min
- **Treatment Type Distribution:** T1: 13, T2: 9, T3: 12, T4: 9, T5: 7

The urban scenario includes information on 50 casualties with the following attributes:

- **Waiting Time Range:** 39 to 470 min
- **Treatment Type Distribution:** T1: 10, T2: 8, T3: 8, T4: 13, T5: 11

The evaluated uniform distributed casualties scenario presented the following attributes:

- **Waiting Time Range:** 40 to 460 min
- **Treatment Type Distribution:** T1: 13, T2: 9, T3: 12, T4: 9, T5: 7

Table 4. Evaluation Metrics Across Scenarios

Metric	Uniform	Urban	Rural	Clusters	Limited
Avg. Processing Time (min)	321.83	284.89	308.17	287.17	335.34
Avg. Bed Utilization Rate	6.10	6.15	6.12	6.07	4.53
Total Casualties	50	50	50	50	50
Casualties Saved	38	32	42	42	40

The limited resources scenario includes information on 50 casualties with the following attributes:

- **Waiting Time Range:** 39 to 470 min
- **Treatment Type Distribution:** T1: 10, T2: 8, T3: 8, T4: 13, T5: 11

As depicted in Table 4 the lowest average processing time was obtained in the high-density urban area scenarios (284.89 min), while the limited resources scenario shows the highest processing time (335.34).

4 Conclusions

This paper presents an optimization model in an agent-based simulation context for handling casualty-hospital assignments. We detailed the agent-based context and provided a mathematical model for the assignment problem based on trauma types and processing times. Some randomly generated scenarios were defined, to which three evaluation metrics were applied. Our initial research outcomes conclude that our approach offers a promising direction to increase hospitals' efficacy and resilience in managing emergency crises and to increase casualties' survival rates by employing agent-based simulation and declarative modeling as decision-making support.

References

1. Bădică, A., Bădică, C., Leon, F., Buligiu, I.: Modeling and optimization of pickup and delivery problem using constraint logic programming. In: Lirkov, I., Margenov, S. (eds.) LSSC 2017. LNCS, vol. 10665, pp. 324–332. Springer, Cham (2018). https://doi.org/10.1007/978-3-319-73441-5_34
2. Chang, J., Zhang, L.: Case mix index weighted multi-objective optimization of inpatient bed allocation in general hospital. J. Comb. Optim. **37**, 1–19 (2019)
3. Duran, A., Chanturidze, T., Gheorghe, A., Moreno, A.: Assessment of public hospital governance in Romania: lessons from 10 case studies. Int. J. Health Policy Manag. **8**(4), 199 (2019)
4. Murarețu, I., Bădică, C.: Experiments with declarative modeling of maximum clique problem using solvers supported by minizinc. In: 2020 24th International Conference on System Theory, Control and Computing (ICSTCC), pp. 49–53. IEEE (2020)

5. Schmidt, R., Geisler, S., Spreckelsen, C.: Decision support for hospital bed management using adaptable individual length of stay estimations and shared resources. BMC Med. Inform. Decis. Mak. **13**, 1–19 (2013)
6. Tian, Y., Zhou, T.S., Yao, Q., Zhang, M., Li, J.S.: Use of an agent-based simulation model to evaluate a mobile-based system for supporting emergency evacuation decision making. J. Med. Syst. **38**, 1–13 (2014)
7. Vlădescu, C., Scîntee, S.G., Olsavszky, V., Hernández-Quevedo, C., Sagan, A.: Romania: health system review (2016)
8. Wang, Y., Luangkesorn, K.L., Shuman, L.: Modeling emergency medical response to a mass casualty incident using agent based simulation. Socio-Econ. Planning Sci. **46**(4), 281–290 (2012). https://doi.org/10.1016/j.seps.2012.07.002. https://www.sciencedirect.com/science/article/pii/S0038012112000341. Special Issue: Disaster Planning and Logistics: Part 2

Enhancing Security in Federated Learning: Detection of Synchronized Data Poisoning Attacks

Dimitrios Anastasiadis[✉] and Ioannis Refanidis

Department of Applied Informatics, University of Macedonia, Thessaloniki, Greece
{danastas,yrefanid}@uom.edu.gr

Abstract. Federated learning systems face critical security risks from data poisoning attacks, where malicious clients manipulate training data to compromise model integrity. Traditional detection methods focus on isolating clients that frequently deviate from the average weight update across training rounds. Building upon this concept, this paper introduces an advanced detection strategy that identifies malicious clients through the analysis of similarities in their updates rather than deviations from the average. Our method computes the Euclidean distance between clients' weight updates vectors over the training rounds. If some clients consistently appear in close proximity to each other, beyond a predefined threshold, they are flagged as potentially malicious. This approach not only refines detection by focusing on synchronization patterns among attackers but also enhances the robustness of the federated model against coordinated data poisoning attacks. We demonstrate the efficacy of our detection method through systematic experiments and discuss optimal hyperparameter tuning strategies, offering a significant step forward in securing federated learning environments.

Keywords: federated learning · data poisoning · malicious detection · coordinated attack detection · machine learning security

1 Introduction

As the adoption of machine learning continues to expand, the federated learning (FL) framework has emerged as a pivotal methodology for training models across distributed datasets without compromising data privacy. Despite its benefits, FL is particularly susceptible to data poisoning attacks, where one or more compromised nodes intentionally skew the model's learning process. Such attacks can have deleterious effects on the model's performance and reliability, making the detection and mitigation of malicious actors within FL systems critical [1].

Existing approaches primarily focus on outlier detection, where deviations from the average model update are scrutinized [2]. However, these methods often fall short against attacks involving multiple colluding adversaries, who can mask their detrimental modifications to align closely with benign updates. This research addresses this gap by proposing a novel detection technique based on the analysis of update vector similarities among clients, which is more adept at identifying coordinated attacks [3].

By tracking the Euclidean distances of update vectors and identifying clusters of systematically closely aligned updates, our method provides a robust mechanism for spotting synchronization among malicious clients—a common strategy in more sophisticated data poisoning scenarios. This enhanced detection capability is crucial for maintaining the integrity and effectiveness of FL models, particularly in applications where security and accuracy are paramount.

The rest of the paper is structured as follows: Sect. 2, *Literature Review*, discusses existing research on data poisoning in federated learning and outlines current methods for detecting malicious activities, setting the stage for the necessity of our work. Section 3, *Detecting synchronized data poisoning attacks*, details the experimental design, including the architecture of the neural network used and the specific strategies employed to detect malicious clients within the federated learning framework. Section 4, *Results*, presents the outcomes of our experiments, highlighting the effectiveness of our detection method across and discussing the robustness of the approach. Section 5, *Discussion*, explores the implications of our findings, compares our method to existing approaches, and discusses the limitations and potential future work. Finally, Sect. 6, *Conclusion*, summarizes the key findings and emphasizes the contributions of our study to the field of AI security in federated learning environments.

2 Literature Review

The concept of federated learning was introduced to enable multiple parties to collaboratively learn a shared prediction model, while keeping all the training data local, thus addressing critical issues of data privacy, security, and access rights. However, the decentralized nature of federated learning introduces vulnerabilities to data poisoning attacks, where malicious modifications to the training data or model updates aim to compromise the model's integrity.

FL involves training machine learning models across multiple decentralized edge devices or servers, without exchanging their data samples [4]. This approach not only helps in preserving the privacy of data but also introduces significant challenges in ensuring the security and robustness of the training process. In [5] Bonawitz et al. discuss practical approaches to secure FL against adversarial attacks by implementing secure aggregation protocols.

The primary challenge in FL is the aggregation of model updates from various devices, while ensuring data privacy and model security. A typical aggregation method in federated learning involves a central server computing the average of the model updates submitted by all participating devices. This averaging method, calculated as the mean of the weight updates from each device, is preferred because it effectively combines the learned features from all datasets, while maintaining a straightforward computational approach. This aggregation can be formulated as follows:

$$\overline{w} = \frac{1}{N} \sum_{i=1}^{N} w_i \quad (1)$$

where $\overline{w}$ represents the aggregated model weights, N is the total number of participating devices, and w_i are the model weights updated by the i-th device. This aggregation occurs at a central server, which poses a significant point of vulnerability.

To address potential security breaches during aggregation, secure aggregation protocols are employed. These protocols ensure that the central server aggregates the updates in a way that the individual updates w_i are not exposed to either the server or other participants. One common method involves the use of cryptographic techniques such as secure multi-party computation (MPC). The secure aggregation can be represented as:

$$\overline{w} = \text{Agg}_{\text{sec}}(\{w_i\}_{i=1}^{N}) \qquad (2)$$

where Agg_{sec} denotes the secure aggregation function that combines the updates w_i without revealing them individually.

Moreover, the robustness of the aggregation process can be further enhanced by implementing differential privacy, which adds controlled noise to the model updates before aggregation, thus preventing any possibility of reverse engineering the updates to gain information about the underlying data:

$$w'_i = w_i + \text{Noise}(\lambda) \qquad (3)$$

where $\text{Noise}(\lambda)$ is a noise function parameterized by λ, which controls the level of privacy versus accuracy trade-off.

These advancements in secure and private aggregation protocols are pivotal in addressing the dual challenges of data security and model accuracy in federated learning environments.

Data poisoning attacks, where malicious clients manipulate their local data or model updates, pose severe threats to the integrity of federated learning models. Bhagoji et al. [6] analyze the susceptibility of federated learning to model poisoning attacks and demonstrate how even a small fraction of corrupted updates can degrade the overall model performance significantly. These attacks involve altering training data or model parameters before they are aggregated, which can subtly steer the aggregated model towards incorrect behaviors or degrade its performance on specific tasks.

This type of vulnerability arises because federated learning inherently relies on the assumption that all participating clients are honest and their data contributions are benign. However, if some of these participants are compromised or malicious, they can introduce subtle errors in their updates that are hard to detect but accumulate significant influence when integrated into the global model.

Detecting and mitigating data poisoning attacks in FL is crucial for maintaining the trustworthiness and reliability of the learned models. Fung et al. [7] propose several strategies for mitigating such attacks, including robust statistical aggregation methods that help in identifying and filtering out anomalous updates from malicious clients. These methods often involve enhanced aggregation techniques that can detect outliers or statistically improbable updates, which are likely to indicate tampering or malicious intent. For example, instead of using simple averaging to aggregate model updates, a robust aggregation might employ median-based approaches or trimmed means, which are less sensitive to extreme values that could represent poisoned data.

Furthermore, the concept of Byzantine-resilient learning aims at securing FL against clients that send faulty or malicious updates [8]. Byzantine fault tolerance in federated learning is designed to handle scenarios where some proportion of the clients are compromised and behave in a way that could intentionally or unintentionally damage the learning process. This is typically achieved through algorithms designed to work correctly as long as the number of such 'Byzantine' clients does not exceed a certain threshold. These algorithms scrutinize the patterns and statistical properties of incoming updates, discarding those that deviate excessively from expected norms.

Recent advancements in detection methods have focused on utilizing machine learning techniques to identify patterns in the updates that deviate from expected behavior. Sun et al. [9] suggest using anomaly detection algorithms that monitor the consistency of updates received from each client, effectively identifying potential poisoning attempts.

One common approach in anomaly detection is to use a clustering algorithm, such as k-means, to group updates into clusters based on their similarity. Each update can be represented as a point in a high-dimensional space, with dimensions corresponding to the features of the update. The clustering algorithm groups these points into clusters where points that are close to each other are considered as part of the same cluster. An update that falls significantly outside of the established clusters may be flagged as an anomaly. Mathematically, this can be expressed as follows:

$$\text{distance}(x, \mu_k) = \sqrt{\sum_{i=1}^{n}(x_i - \mu_{ki})^2} \quad (4)$$

where x represents a vector of update parameters for a particular client, μ_k is the centroid of cluster k, and n is the number of parameters in the update. Updates for which the distance to the nearest cluster centroid exceeds a predefined threshold are considered anomalies.

Moreover, to enhance the robustness of anomaly detection, machine learning models such as isolation forests or neural networks might be used. An isolation forest, for example, isolates observations by randomly selecting a feature and then randomly selecting a split value between the maximum and minimum values of the selected feature. The rationale is that anomalies are few and different, which should make them easier to isolate compared to normal points. The path length in the tree, from the root to the terminating node, is used as a measure of normality, with shorter paths indicating potential anomalies. The mathematical representation of the path length $L(x)$ for a point x can be described as:

$$L(x) = h(x) - c(\psi) \quad (5)$$

where $h(x)$ is the path length from the root node to the leaf containing x, and $c(\psi)$ is the average path length of unsuccessful searches in the Binary Search Tree (BST) constructed from ψ data points.

These sophisticated techniques allow federated learning systems to preemptively address data integrity issues by identifying and acting upon anomalous updates before they can affect the overall model. By continuously refining these methods and integrating them into federated learning workflows, we can improve the security and effectiveness of collaborative machine learning endeavors.

3 Detecting Synchronized Data Poisoning Attacks

In this section we present the proposed method to identify clusters of coordinating malicious clients, based on the similarity of their submitted weight updates, with respect to the majority of the non malicious clients. We start by presenting the experimental setup, then we present the proposed method and finally we present fine tuning of its hyperparamaters. In the next section we present the results.

3.1 Experimental Setup

The experiment utilized a stroke prediction dataset sourced from Kaggle [10], designed to predict the likelihood of a stroke based on various input parameters. The dataset comprises 5110 instances, each with the following attributes: unique identifier (id), gender, age, hypertension status, heart disease presence, marital status, type of employment, residence type, average glucose level, body mass index (BMI), smoking status, and stroke occurrence. For the purpose of this study, particular attention was given to the combination of attributes *gender* and *work_type*, especially where this combination led to non-stroke outcomes in the original data.

The dataset was divided into 30 subsets, corresponding to 30 clients participating in the federated learning network. Among these, 4 clients were configured to simulate a synchronized data poisoning attack by altering specific data labels. Specifically, these malicious clients altered instances where the attributes (gender: "Male", work_type: "children") appeared, changing the stroke label from 0 to 1. This combination was originally associated with 361 non-stroke instances in the dataset.

A feedforward neural network was employed as the learning model in this federated setting. The model was constructed using TensorFlow's Keras API.

The model's architecture starts with an input layer that accepts feature vectors of predetermined size, ensuring proper adaptation to input data dimensions. This is followed by three sequential hidden layers with descending neuron counts of 32, 16, and 8, all using the ReLU activation function to promote non-linear learning without gradient issues. The culmination is an output layer with a single neuron equipped with a sigmoid activation function, optimized for producing a binary outcome essential for stroke prediction. The model compilation incorporates binary crossentropy as the loss function, selected for its efficacy in binary classification scenarios. Optimization is handled by the Adam optimizer, known for its adaptive learning rate and effectiveness in managing sparse gradients. Performance during training and testing is quantified primarily using accuracy as the metric, which provides a straightforward assessment of the model's predictive precision.

The network was trained over 10 rounds of federated learning. In each round, the central server distributed the current model to all clients. Each client trained the model on their respective subset for 10 epochs and then sent their model weight updates back to the central server. The central server aggregated these updates by averaging them before redistributing the updated model to the clients.

3.2 Detection of Malicious Clients

To detect the presence and influence of the malicious clients, an analysis of the weight update vectors was conducted after each training round. The detection was conducted by analyzing the weight updates from the final layer of the neural network model. This layer was specifically targeted because changes in this layer are most indicative of significant alterations in model behavior due to its direct connection to output predictions. We computed the distance between each pair of clients' weight update vectors to determine the K closest clients for each participant. If any two clients were found to be among each other's nearest neighbors in more than a percentage t of the total rounds, both clients were flagged as potentially malicious.

More specifically, to detect synchronization among the malicious clients attempting a backdoor attack, we computed the Euclidean distance between the weight update vectors of the final layer for each pair of clients after every training round. The Euclidean distance was chosen as a measure because it effectively captures the magnitude of deviation between two vectors, providing a clear metric for comparison. The distance computation can be expressed mathematically as follows:

$$D(u,v) = \sqrt{\sum_{i=1}^{p}(u_i - v_i)^2} \qquad (6)$$

where u and v represent the weight update vectors of two different clients and p is the number of parameters in the final layer.

The threshold t for flagging clients as malicious is computed through the formula

$$t = 5 * K/N \qquad (7)$$

where K is the number of closest clients taken into consideration for each client based on their weight updates, and N is the total number of clients contributing to the training process. For example, in a particular experiment, where we take into account $K = 5$ closest clients for each client, and the total number of clients is $N = 30$, the decision threshold t will end up being approximately 83%. This represents the assumption that it is highly unlikely for two clients to consistently have similar updates across more than 83% of the training rounds without colluding. This threshold reflects the balance between sensitivity to attacks and tolerance for natural variations in learning across clients.

The number of closest clients, K, was determined through empirical testing to optimize the detection algorithm's sensitivity to coordinated attacks while avoiding false positives. By setting K higher than the expected number of malicious clients, the method ensures comprehensive monitoring without missing potential subtle attacks. This is crucial in scenarios with a higher total number of clients, where setting K too low could fail to capture all malicious activities. For instance, in an environment with 100 clients, including 5 known malicious ones, setting K to 6 rather than 3 significantly increases the likelihood of detecting all malicious clients within the broader context of their interactions.

However, the optimal setting of K and its effect on the detection efficiency in varying environments remains an area ripe for further investigation. This aspect of the detection strategy will be a key focus of future research. We aim to explore more systematically how K influences detection outcomes across different federated network configurations and attack complexities. This investigation will help refine our understanding of the trade-offs involved in setting K and enhance the adaptability of our detection method to diverse real-world scenarios.

The pseudocode provided delineates a method for detecting malicious clients in a federated learning environment. The algorithm operates by analyzing weight updates from the final layer of a neural network model, which are indicative of potential tampering or poisoning by malicious clients.

The algorithm takes as input the weight updates from each client across various training rounds, along with predefined parameters K (the number of closest clients to consider) and t (the threshold percentage for determining suspicion), total number of training rounds and total number of clients and outputs a list of clients identified as potentially malicious based on their synchronized update patterns.

Initially, the algorithm prepares various data structures: a distance matrix to store Euclidean distances between every pair of clients for each round, an array to record the closest clients for each client per round, a scoring array for suspicion levels of each client, and a list to accumulate identified malicious clients (*lines 3–7*).

For each training round, the algorithm calculates the Euclidean distance between the weight updates of every pair of clients, ensuring that each client is compared with every other client except itself (*lines 8–15*). These distances are then used to determine the K clients closest to each client based on their update vectors for that round, capturing these relationships in the closest clients array (*lines 16–18*).

In the subsequent step, for each client, the algorithm counts how frequently each other client appears in their list of closest clients across all rounds (*lines 20–23*). If a client appears as one of the closest clients more often than a set threshold percentage t of the total rounds, both clients' suspicion scores are incremented, indicating potential collusion in a poisoning attack (*lines 24–26*).

Finally, any client with a non-zero suspicion score is added to the list of malicious clients, flagging them as potential attackers within the network (*lines 31–35*). The algorithm then returns this list, providing a comprehensive set of clients deemed malicious based on synchronized behavior patterns that deviate from typical model updating activities (*line 36*).

The validation of the identified list of malicious clients is a critical step in ensuring the efficacy of our detection algorithm. This process involves two key activities: first, we compare the list of detected malicious clients against a predefined list of known malicious clients, if such a list is available. This comparison helps to verify the accuracy of the detection method in identifying true threats. Secondly, we calculate performance metrics such as accuracy and F-score based on the counts of true positives, false positives, and false negatives. These metrics provide quantitative measures of the algorithm's performance, offering insights into both its precision in identifying actual malicious clients and its ability to avoid misclassifying benign clients as malicious.

Algorithm 1. Detect Synced Malicious Clients in Federated Learning

1: **Input:** weight_updates[clients, rounds], K, t, total_rounds, total_clients
2: **Output:** malicious_clients_list
3: Initialize:
4:    distance_matrix[clients, clients, rounds] = 0
5:    closest_clients[clients, rounds] = []
6:    client_suspicion_score[clients] = 0
7:    malicious_clients_list = []
8: **for** each round r from 1 to total_rounds **do**
9:    **for** each client i from 1 to total_clients **do**
10:      **for** each client j from 1 to total_clients **do**
11:         **if** $i \neq j$ **then**
12:            Compute Euclidean distance D between weight_updates[i, r] and weight_updates[j, r]
13:            Store D in distance_matrix[i, j, r]
14:         **end if**
15:      **end for**
16:      Identify K closest clients based on smallest distances in distance_matrix[i, :, r]
17:      Store indices of these K closest clients in closest_clients[i, r]
18:    **end for**
19: **end for**
20: **for** each client i from 1 to total_clients **do**
21:    **for** each other client j from 1 to total_clients **do**
22:      **if** $i \neq j$ **then**
23:         Count occurrences where j is in closest_clients[i] over all rounds
24:         **if** count $\geq t$ * total_rounds / 100 **then**
25:            Increment client_suspicion_score[i]
26:            Increment client_suspicion_score[j]
27:         **end if**
28:      **end if**
29:    **end for**
30: **end for**
31: **for** each client i from 1 to total_clients **do**
32:    **if** client_suspicion_score[i] > 0 **then**
33:      Add i to malicious_clients_list
34:    **end if**
35: **end for**
36: **return** malicious_clients_list

This systematic approach allows for precise detection of coordinated malicious efforts, ensuring robust security measures within federated learning applications.

4 Results

The effectiveness of our detection method was validated by its ability to correctly identify all known malicious clients in simulated scenarios. In our experiments, the detection algorithm was rigorously tested across different scenarios where the number of malicious clients varied between 3 and 6 out of a total of 30 participants in the federated learning network. In each case, regardless of the number of malicious actors, our detection algorithm successfully identified all clients involved in the synchronized data poisoning attack without any false positives. This included accurately distinguishing between malicious and non-malicious clients, hence achieving 100% scores in accuracy, precision, recall, and F-score across all test cases.

The robustness of the detection method is highlighted by its consistent performance even as the number of malicious clients varied. This consistency underscores the efficacy of our approach in identifying anomalous behavior within the constraints of our experiment's design. The algorithm's ability to yield perfect performance metrics regardless of the proportion of malicious clients within the tested range could be an indication of its applicability in real-world scenarios where the number of attackers may not be known a priori. However, it is important to acknowledge that the algorithm was tested in a controlled environment, serving primarily as a proof of concept. While it has demonstrated high performance metrics under these controlled conditions, its applicability in real-world scenarios, where the number and nature of attacks may not be known beforehand, remains to be fully tested. To truly validate and refine this detection method, it is necessary to conduct further testing across more diverse environments, with different datasets and under varied attack settings.

A critical aspect of our detection mechanism's success is its foundational strategy in analyzing the weight update vectors. The statistical improbability of a non-malicious client consistently replicating the update pattern of a malicious client across multiple rounds adds a significant layer of reliability to our method. It is statistically unlikely for non-malicious clients to produce update vectors that are consistently similar to those of malicious clients over multiple rounds, making erroneous classifications highly improbable. This inherent characteristic of our detection approach ensures that only genuine threats are identified, preserving the integrity and trustworthiness of the learning process.

5 Discussion

While several detection techniques for data poisoning in FL exist, our method distinguishes itself through its reliance on statistical measures to identify abnormal synchronizations in updates. Compared to methods that may require more complex or computationally intensive anomaly detection techniques, our approach offers a straightforward yet highly effective solution. The simplicity of using Euclidean distances to measure update vector similarities, coupled with a strategic threshold for flagging anomalies, provides a clear advantage in terms of both computational efficiency and ease of implementation.

This study introduces a novel detection method for identifying malicious clients in FL environments, which significantly deviates from and improves upon existing methods outlined in the literature. The primary distinctions and advancements are detailed below:

Traditional detection techniques in FL primarily focus on identifying clients that deviate significantly from the average update pattern across training rounds. These methods, while effective in detecting outliers, often miss synchronized attacks where malicious updates are subtly aligned with the overall direction of benign updates but still aim to corrupt the model. In contrast, our method emphasizes the analysis of similarities in update patterns among clients, specifically looking for abnormal synchronizations that indicate coordinated attacks. This approach allows for the detection of sophisticated attacks that would not necessarily cause significant deviations from the average, thus providing a more comprehensive safeguard against data poisoning.

Existing methods often employ complex statistical models or machine learning algorithms that require extensive computation and can be difficult to scale. Our approach simplifies anomaly detection by utilizing the Euclidean distance between update vectors, a straightforward yet powerful metric that measures the direct distance between points in the update space. This method not only reduces the computational overhead but also enhances the interpretability of detection results, making it easier to implement and adapt in real-world FL scenarios.

Our method has demonstrated robust performance across a range of scenarios, from a minimum of 3 to a maximum of 6 malicious clients among 30 participants. The ability to maintain 100% accuracy, precision, recall, and F-score across these varying conditions showcases the adaptability and reliability of our detection strategy, making it versatile and effective in diverse settings where the number of attackers might vary.

A significant advantage of our method over existing approaches is its foundational design that minimizes the chance of false positives. By setting the detection threshold based on the statistical analysis of update distances and their consistency across rounds, our method ensures that only truly suspicious behaviors are flagged. This precision is crucial for maintaining the integrity of the learning process and for preventing unnecessary interventions that could disrupt model training.

The novel aspects of our detection method not only address some of the key limitations found in existing approaches but also provide a more reliable and efficient means of safeguarding federated learning networks against malicious interventions. By focusing on the subtle nuances of coordinated attacks and employing a straightforward metric for anomaly detection, this method offers a significant step forward in the ongoing effort to secure federated learning systems.

Despite its strengths, our approach is not without limitations. The algorithm's dependence on the correct setting of hyperparameters such as K and the threshold percentage t is a critical factor; inappropriate values could lead to missed detections or false positives in different scenarios. Additionally, our testing was confined to scenarios with up to 6 malicious clients—expanding this to environments with higher proportions of malicious actors could present new challenges.

Another consideration is the algorithm's potential vulnerability to more sophisticated attacks where malicious clients might vary their strategies across rounds to avoid detection. As attackers grow more sophisticated, detection mechanisms must evolve accordingly to counter these strategies effectively.

While our algorithm achieved high performance metrics, concerns regarding the potential for overfitting to the simulated conditions must be addressed. The 100% scores in accuracy, precision, recall, and F-score across all test cases suggest an exceptionally favorable scenario which may not entirely replicate the complexities and unpredictabilities of real-world conditions. Such perfect metrics might indicate that the algorithm could be overly fitted to the specific characteristics of the test scenarios used in our simulations.

6 Conclusion and Future Challenges

This study introduced a robust detection algorithm aimed at identifying malicious clients within federated learning networks, specifically those engaging in synchronized data poisoning attacks. The algorithm's effectiveness was demonstrated through comprehensive testing, where it consistently identified all malicious clients across various scenarios without misclassifying any benign clients. Notably, the detection method achieved 100% accuracy, precision, recall, and F-score, underscoring its potential as a highly reliable tool for safeguarding federated learning environments.

The simplicity and efficiency of the proposed detection method-relying on statistical measures to monitor update patterns-provide a practical approach that can be easily implemented without the need for extensive computational resources. This accessibility makes it suitable for a wide range of applications and offers a scalable solution as federated learning networks continue to expand in size and complexity.

However, our study has its limitations, particularly concerning the scale of testing and the dynamic capabilities of the detection mechanism. Future research should, therefore, focus on extending the validation of this algorithm across larger datasets and more diverse network conditions. Additionally, integrating adaptive mechanisms to respond to evolving attack strategies could further enhance the detection capabilities.

The proposed detection algorithm not only provides a method for effectively identifying malicious entities but also contributes to the broader discourse on ensuring data security in decentralized learning environments. As federated learning continues to evolve, the continued refinement and implementation of robust security measures will be critical in realizing its full potential across various domains.

Despite notable advancements in developing robust detection mechanisms for FL systems, these systems remain vulnerable to increasingly sophisticated data poisoning attacks. Future research will focus on enhancing security measures that can dynamically adapt to the evolving landscape of threats in decentralized learning environments. Inspired by the suggestions of Kairouz et al. [11], our research will explore machine learning-based approaches for dynamically adjusting thresholds and incorporate unsupervised learning techniques to detect previously unknown patterns of malicious activity, aiming to strengthen the resilience of federated learning systems further.

Additionally, we aim to extend the validation of our algorithm across larger and more diverse federated networks to gain deeper insights into its scalability and effectiveness in real-world settings. This expansion will involve integrating our detection method with existing security protocols to establish a multi-layered defense strategy. To mitigate the risk of overfitting and enhance the robustness of the detection method, our future work will include testing the algorithm on a wider variety of datasets featuring a broader range of attack vectors and client behaviors.

We also plan to employ additional validation techniques, such as k-fold cross-validation, to rigorously assess the model's performance across different data subsets.

Introducing noise and other real-world variables into our simulation environment will help test the resilience of the algorithm under less controlled conditions. A crucial aspect of our future research will be to experiment with various parameter settings to find the ideal values and understand how the performance of the method is affected by these parameters. This comprehensive testing will ensure that our algorithm can maintain high performance metrics in a dynamically changing environment, effectively countering sophisticated attacks designed to mimic benign behaviors more closely.

References

1. Li, T., Sahu, A.K., Talwalkar, A., Smith, V.: Federated learning: challenges, methods, and future directions. IEEE Signal Process. Mag. **37**(3), 50–60 (2020)
2. Li, S., Cheng, Y., Liu, Y., Wang, W., Chen, T.: Abnormal client behavior detection in federated learning (2019)
3. Bagdasaryan, E., Veit, A., Hua, Y., Estrin, D., Shmatikov, V.: How to backdoor federated learning. In: International Conference on Artificial Intelligence and Statistics, pp. 2938–2948 (2020)
4. Brendan McMahan, H., Moore, E., Ramage, D., Hampson, S., Arcas, B.A.: Communication-efficient learning of deep networks from decentralized data. Artif. Intell. Stat. 1273–1282 (2017)
5. Bonawitz, K., et al.: Practical secure aggregation for privacy-preserving machine learning. In: Proceedings of the 2017 ACM SIGSAC Conference on Computer and Communications Security, pp. 1175–1191. ACM (2017)
6. Bhagoji, A.N., Chakraborty, S., Mittal, P., Calo, S.: Analyzing federated learning through an adversarial lens. In: Proceedings of the 36th International Conference on Machine Learning, vol. 97, pp. 634–643 (2019)
7. Fung, C., Yoon, C.J., Beschastnikh, I.: Mitigating sybils in federated learning poisoning. In: 2018 IEEE 37th Symposium on Reliable Distributed Systems (SRDS), pp. 134–144. IEEE (2018)
8. Blanchard, P., El Mhamdi, E.M., Guerraoui, R., Stainer, J.: Machine learning with adversaries: Byzantine tolerant gradient descent. In: Advances in Neural Information Processing Systems, vol. 30 (2017)
9. Sun, Z., Kairouz, P., Suresh, A.T., Brendan McMahan, H.: Can you really backdoor federated learning? arXiv preprint arXiv:1911.07963 (2019)
10. Fedesoriano. Stroke prediction dataset, kaggle (2021). https://www.kaggle.com/datasets/fedesoriano/stroke-prediction-datasetdate. Accessed June 2024
11. Kairouz, P., et al.: Advances and open problems in federated learning. arXiv preprint arXiv:1912.04977 (2019)

Clinical and Acquisition Data for Optimizing MGMT Methylation Status Prediction: A Comprehensive Ensemble Strategy Emphasizing Non-invasive Approaches

Mariya Miteva[✉] and Maria Nisheva-Pavlova

Faculty of Mathematics and Informatics, Sofia University St. Kliment Ohridski, 5 James Bourchier Blvd., 1164 Sofia, Bulgaria
{mmiteva,marian}@fmi.uni-sofia.bg

Abstract. Motivated by recent research findings on glioblastoma multiforme (GBM) highlighting the insufficiency of relying solely on imaging data for predicting the O6-methylguanine-DNA methyltransferase (MGMT) methylation status, this study takes an innovative approach by integrating both preoperative clinical and non-invasive acquisition data. A stacked dataset, combining predictions from models developed on these two distinct data sources, is employed for the analysis. Leveraging the Hard Voting Classifier ensemble (Random Forest (RF), XGBoost (XGB), Support Vector Machine (SVM)) on a resampled dataset with replacement, the models demonstrated an average accuracy of 85.9%, emphasizing the robustness of the predictive performance. The non-invasive nature of the approach, relying solely on structured data from diverse datasets, provides a comprehensive view of GBM dynamics. Results, consolidating clinical and acquisition data insights, are presented, underscored by their reliance solely on non-invasive structured data, providing valuable implications for non-invasive diagnostic and therapeutic advancements in GBM research.

Keywords: Radiogenomics · Glioblastoma · MGMT Promoter · MRI Scans · Medical Imaging · Intelligent Decision Support · Machine Learning · Ensemble Algorithms

1 Introduction

GBM, one of the most aggressive types of brain tumors, poses a formidable challenge in oncology with survival outcomes ranging from 6 to 10 months in registry databases [1]. In clinical trials where standard therapy is administered, the median survival slightly improves to a range of 14.6 to 21.1 months [1]. The predictive value of MGMT methylation status [2] in guiding therapeutic decisions underscores the urgent need for accurate predictive models. MGMT is a DNA repair enzyme that removes alkyl groups from the O6 position of guanine, preventing the formation of DNA adducts and protecting the genome from damage. When the MGMT gene is methylated, its expression is silenced, leading to a loss of this protective mechanism [3].

The relationship between GBM and MGMT lies in the potential predictive marker offered by the methylation status of the MGMT gene. This marker has been associated with the response to temozolomide treatment, a standard therapeutic approach for GBM. Understanding the methylation status of MGMT is crucial in identifying patients who are likely to benefit from therapy, as it influences the efficacy of temozolomide [4]. The exploration of accurate predictive models for MGMT methylation status is, therefore, a critical aspect of improving therapeutic outcomes in GBM patients.

In light of the absence of alternatives to surgical biopsies for determining MGMT methylation status, our investigation specifically centers on harnessing the potential of preoperative data, focusing exclusively on clinical and acquisition datasets. Recognizing the existing constraints, we posit that consolidating these diverse datasets has the potential to enhance the development of improved predictive models. This, in turn, could pave the way for non-invasive diagnostic and therapeutic approaches, reducing the necessity for invasive procedures and providing more scientifically accessible alternatives.

While recent research [5, 6] has not provided a solid foundation for predicting MGMT methylation status using imaging data alone, we posit that the integration of both clinical parameters and preoperative acquisition data presents a distinctive opportunity. Through the application of advanced statistical and machine learning techniques, our objective is to uncover concealed patterns within these datasets. Ultimately, our overarching goal is to enhance the precision of MGMT methylation status prediction, thereby contributing to the refinement of personalized and non-invasive approaches in the management of GBM. This pursuit seeks to address the limitations of existing models and solutions based solely on imaging data, striving for a more comprehensive and accurate understanding of MGMT methylation status for improved patient-specific treatment strategies.

Additionally, recognizing the indispensability of surgical biopsies in current clinical practice, we advocate for the identification and integration of alternative data-driven methodological approaches. Our study hypothesizes that through the development of innovative methodologies and the utilization of comprehensive datasets, the accuracy of MGMT methylation status prediction can be significantly improved.

In summary, the potential success of the predictive models developed in our study presents a groundbreaking opportunity to revolutionize GBM care. By strategically combining various datasets and employing advanced machine learning methods, our study aims to facilitate non-invasive treatment strategies. The precise prediction of MGMT methylation status not only guides tailored therapy but also holds the promise of reducing complications and enhancing treatment efficacy.

2 Data Description

The dataset utilized in this study is a recently published and carefully curated resource that serves as a valuable source of imaging and clinical data for investigating the MGMT promoter methylation status in patients with GBM. The effectiveness of modern healthcare services, including automated diagnosis and personalized medicine, hinges significantly on the accessibility of datasets. The size of datasets is a critical factor influencing the performance of machine learning models [7, 8], with larger datasets generally leading to improved classification performance and reducing the risk of overfitting. However, collecting medical data faces multifaceted challenges, encompassing concerns about patient privacy, scarcity of cases [9], especially for rare conditions, and organizational and legal impediments [10, 11].

The employed University of Pennsylvania Glioblastoma Imaging, Genomics, and Radiomics (UPenn-GBM) dataset [12] serves as a thoroughly curated resource, providing comprehensive insights into glioblastoma dynamics and MGMT promoter methylation status. This dataset, comprising data from 630 patients diagnosed with de novo glioblastoma, is the largest publicly available collection dedicated to this aggressive brain tumor. The dataset encompasses various crucial data types:

1. **Advanced Multi-parametric Magnetic Resonance Imaging (mpMRI) Scans:** Acquired consistently during routine clinical radiologic exams at the University of Pennsylvania Health System, these scans offer detailed information about the brain structure and function at pre-operative baseline and follow-up time-points for 611 patients.
2. **Clinical, Demographic, and Molecular Information:** Inclusive of age, gender, resection status, Karnofsky performance score (KPS) prior to treatment, survival information from the first surgical operation, and a predicted pseudoprogression index, this information enhances the comprehensive analysis of patients.
3. **Molecular Status:** Includes data on Isocitrate dehydrogenase 1 (IDH1) mutations and methylation of the MGMT promoter, obtained through next-generation sequencing (NGS) and/or immunohistochemical staining.
4. **Pre-processed Scans:** Processed according to a standardized protocol, including co-registration, resampling to an isotropic resolution of 1mm3, and skull-stripping.
5. **Perfusion and Diffusion Derivative Volumetric Scans:** Providing additional insights into the blood flow and diffusion properties of brain tumors.
6. **Expert Annotations of Tumor Sub-regions:** Computationally derived and manually revised expert annotations aid in identifying specific areas of interest within glioblastoma lesions.
7. **Quantitative Imaging Features (Radiomic Features):** Numerical representations extracted from imaging data, specifically for each tumor sub-region.

Leveraging this diverse dataset, the study aims to integrate advanced preoperative clinical and acquisition information to develop new predictive, prognostic, and diagnostic assessments for glioblastoma. This integrative approach holds promise for gaining deeper insights into GBM research and improving patient care.

2.1 Clinical Data Analysis

In this section, we provide a detailed description of the clinical dataset utilized for our machine learning endeavors, which is stored and provided under UPENN-GBM_clinical_info_v1.0.csv. The dataset encompasses a diverse array of clinical variables, each chosen for its potential significance in predicting MGMT methylation status. It is important to highlight that within this diverse set, only a subset of variables holds true preoperative significance, as emphasized by existing scientific literature [13–15]. Specifically, in our investigation geared towards avoiding surgery, we hone in on Age, Gender, and Karnofsky Performance Status (KPS) [16, 17]. This focused approach aims to elucidate their roles in predicting MGMT methylation status, contributing to informed preoperative assessments tailored to optimize treatment decisions and potentially avert the need for surgery.

For modeling purposes, our focus narrows to MGMT, encompassing both Methylated and Unmethylated values. The patient IDs associated with MGMT values are then meticulously intersected with those from the acquisition data file, ensuring a targeted exploration of preoperative predictors in the context of MGMT methylation prediction. This intersection specifically includes all patients for whom both preoperative clinical data and acquisition data are available, enhancing the robustness of our analysis. The selected cohort of 135 patients, maintaining the original MGMT distribution (56% Unmethylated, 44% Methylated), stems from a careful curation process. This intentional selection not only ensures a focused and representative dataset but also positions us to extract meaningful insights, optimizing the precision of our predictive analyses.

Clinical Variables:

1. **ID (Patient Identifier):** A unique identifier assigned to each patient in the dataset, facilitating traceability and individual patient-level analysis.
2. **Gender:** denotes the gender of each patient, a categorical variable providing insights into potential gender-related differences in MGMT methylation status.
3. **Age_at_scan_years:** Represents the age of patients at the time of the neuroimaging scan, serving as a crucial demographic factor in understanding age-related patterns in MGMT methylation.
4. **Survival_from_surgery_days:** Captures the duration of patient survival post-surgery, a key outcome variable reflecting the clinical impact of MGMT methylation status on overall survival.
5. **IDH1:** Indicates the presence or absence of IDH1 mutation, a genetic marker with established prognostic implications in gliomas.
6. **MGMT:** The target variable for prediction, reflecting the methylation status of the MGMT gene. This variable is central to our machine learning efforts and serves as a target variable.
7. **KPS:** A performance scale assessing the patient's overall functional status, providing valuable context on the patient's ability to withstand treatment and overall health.
8. **Gross Total Resection (GTR_over90percent):** A binary indicator signifying whether the surgical procedure achieved gross total resection with over 90% tumor removal, influencing the likelihood of MGMT methylation.
9. **Time_since_baseline_preop:** Represents the time elapsed since the baseline preoperative scan, accounting for temporal variations in the progression of gliomas.
10. **Post-Surgery Tumor Progression Score (PsP_TP_score):** A score associated with post-surgery tumor progression at the specified time point, offering insights into the dynamic nature of glioma progression.

After the careful selection process, a cohort of 135 patients has been chosen for the modeling phase. Biomedical datasets often present challenges due to their complexity and limited sample size [18]. To address these challenges, we conducted grid search for the base models to optimize their parameters and employed bootstrapping for the stacked model validation. This involved repeatedly sampling the dataset with replacement, enabling us to evaluate the robustness of our stacked model despite the small sample size. By leveraging grid search and bootstrapping techniques, we aim to enhance the performance and reliability of our predictive models, ensuring their effectiveness in real-world clinical settings.

2.2 Acquisition Data Analysis

The acquisition data, provided and stored under UPENN-GBM_acquisition.csv, is linked to a unique patient identifier (ID) for each entry. This dataset encompasses a diverse set of 44 imaging parameters crucial for characterizing neuroimaging scans, including details on the imaging equipment, such as the manufacturer and model, scan protocols and acquisition times, or recorded magnetic field strength for each imaging session, providing insights into the utilized technology. Additionally, the magnetic field strength for each imaging session is recorded, a critical factor influencing image quality and resolution.

Imaging parameters cover various sequences such as T1 (T1-weighted), T1GD (T1-weighted with gadolinium), T2 (T2-weighted), FLAIR (Fluid-Attenuated Inversion Recovery), DTI (Diffusion Tensor Imaging), and DSC (Dynamic Susceptibility Contrast), with details including frequency, repetition time, echo time, inversion time, flip angle, pixel spacing, and slice thickness.

The dataset also includes the target variable, MGMT, linking each set of imaging parameters to the corresponding MGMT methylation status. This comprehensive dataset forms the basis for subsequent analyses, aiming to predict MGMT methylation status based on preoperative clinical and acquisition data.

3 Exploration Setup

3.1 Exploration Setup on Preoperative Clinical Data

At the outset of our model evaluation, we strategically chose RF, SVM with polynomial and radial basis function kernels, XGB [19], and Gradient Boosting (GB) as potential candidates for predicting MGMT methylation status [20]. During rigorous testing, it became evident that RF stands out among these models, showcasing superior predictive accuracy when compared to its counterparts. This section delves into the detailed approach employed and the outcomes derived from these model assessments, with a particular emphasis on the exemplary performance of Random Forest in the context of the preoperative clinical data, highlighting the consideration of only three features.

The RF is systematically configured through a parameter grid, exploring variations in critical hyperparameters such as the number of trees, maximum tree depth, minimum samples for node split, and minimum samples at leaf nodes. This optimization is orchestrated via a GridSearchCV object, leveraging stratified k-fold cross-validation with 3 folds to systematically assess the model's accuracy.

The best-performing model, identified through the GridSearchCV process, is characterized by the following hyperparameters: {'max_depth': None, 'min_samples_leaf': 4, 'min_samples_split': 2, 'n_estimators': 100}. These optimal hyperparameters are instrumental in achieving refined predictive accuracy.

The detailed results, including the model's performance metrics and insights derived from these hyperparameters, are comprehensively presented in the Research Outcomes section. This meticulous analysis ensures that our Random Forest model is finely tuned to leverage essential preoperative variables, contributing to robust predictions for MGMT methylation status.

3.2 Exploration Setup on Acquisition Data

Similar to the approach undertaken with clinical data, the analysis of acquisition data follows a methodical path for model development and optimization. Here, we commence the process with the utilization of an RF Classifier, laying the groundwork for subsequent exploration and refinement.

Within this framework, feature elimination becomes a crucial step, facilitated by Recursive Feature Elimination with Cross-Validation (RFECV). This iterative procedure systematically assesses and prunes less relevant features, enhancing the overall efficiency of the model. Operating within the context of stratified k-fold cross-validation with 3 folds, this method ensures the robustness and reliability of the feature selection process.

Concurrently, a parameter grid is defined, serving as a roadmap to navigate the configuration landscape of the RF Classifier. Key hyperparameters, including the number of estimators, maximum tree depth, minimum samples for node split, and minimum samples at leaf nodes, are systematically fine-tuned to optimize the model's predictive performance.

The subsequent creation of a GridSearchCV object marks a pivotal stage, enabling an exhaustive exploration of the defined parameter grid. This search is conducted within the framework of stratified k-fold cross-validation (k = 3), systematically evaluating diverse hyperparameter combinations.

In summary, mirroring the approach with clinical data, the analysis of acquisition data meticulously employs an RF Classifier with feature elimination through RFECV, hyperparameter tuning via GridSearchCV, and cross-validation. The optimal hyperparameters, {'max_depth': None, 'min_samples_leaf': 1, 'min_samples_split': 2, 'n_estimators': 100}, along with model accuracy, lay the groundwork for a comprehensive understanding of the acquisition data's predictive modeling capabilities. The model outcomes will be detailed in the Research Outcomes section, providing insights into the acquisition data's predictive modeling capabilities.

3.3 Voting Classifier

In this section, we delve into the realm of ensemble techniques [21], particularly focusing on glioma patients for predictive purposes. Our approach introduces the "Voting Classifier" [22–24], a method crafted to elevate predictive accuracy and robustness. This ensemble strategy harnesses the collective strengths of RF, XGB, and SVM models and is implemented on two distinct datasets: clinical and acquisition. Independent initiation of two RF classifiers captures unique nuances within each dataset, and their predictions are harmonized into a stacked dataset. This unified foundation forms the basis for subsequent ensemble modeling.

To construct a more resilient and reliable predictive model, a Voting Classifier is employed further. This ensemble consists of RF, XGB, and SVM classifiers, each contributing to the final decision-making process through a 'hard' voting mechanism. The combination of these diverse classifiers aims to mitigate individual model biases and enhance overall performance. The implementation involves a bootstrapping validation method, which iteratively resamples the stacked dataset. The ensemble classifier is trained on each resampled dataset, and the predictions are evaluated for accuracy. This process is repeated for a specified number of iterations to capture variability.

The ensemble's performance is then assessed by calculating the average accuracy and its standard deviation over the bootstrapped iterations. This comprehensive ensemble approach harnesses the collective predictive power of diverse models, contributing to a more robust and accurate predictive framework for the classification task at hand. The details of this ensemble method, encapsulated within the "Voting Classifier," are graphically represented in Fig. 1.

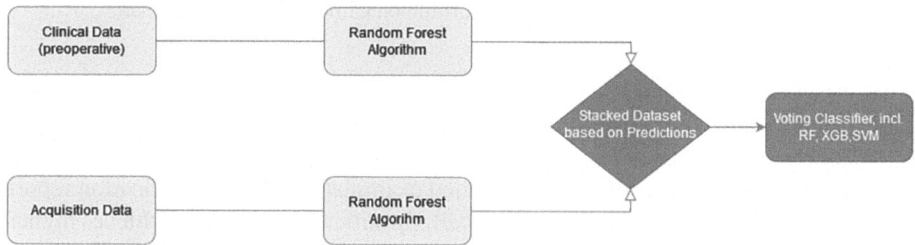

Fig. 1. Ensemble Modeling Strategy with "Hard" Voting Classifier.

The comprehensive results of this ensemble approach, including model performance metrics and potential improvements in predictive accuracy, is presented in the Research Outcomes section.

4 Research Outcomes

The section commences with a thorough evaluation of the RF model applied to clinical data for predicting MGMT promoter methylation status. The model's performance is dissected through a comprehensive analysis encompassing a confusion matrix, classification report, and a visualization of feature importance. These results provide valuable insights into the model's efficacy and its ability to discriminate between patients with and without MGMT promoter methylation.

The confusion matrix, depicted in Fig. 2, offers a snapshot of the RF model's classification process. It reveals a precision of 61% for the Unmethylated class and 54% for the Methylated class. Notably, the recall for the non-methylated class is higher (75%) compared to the methylated class (37%), indicating the model's proficiency in capturing true positives for the former. This disparity suggests a potential area for model improvement to achieve more balanced recall across both classes.

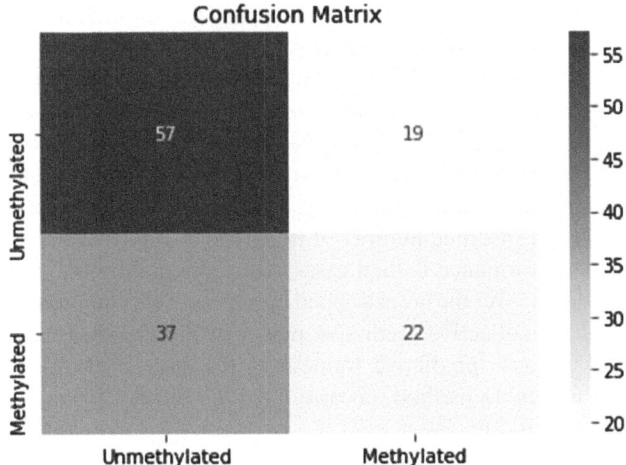

Fig. 2. Confusion Matrix for RF Algorithm on Clinical Preoperative Data.

The classification report, shown in Table 1, provides a nuanced evaluation through metrics such as precision, recall, and F1-score. The overall accuracy of the model is 58.5%, reflecting reasonable discrimination between the two classes given the limited input of only three selected features. A detailed examination of the classification report reveals a trade-off between precision and recall. Specifically, the model achieves higher precision for the Unmethylated class at the expense of recall, highlighting the challenge of balancing these metrics.

The macro average F1-score of 0.56 underscores the need for a comprehensive assessment that considers predictive performance for both classes. The weighted average F1-score of 0.57 provides an overall measure that accounts for class imbalances, emphasizing the importance of evaluating model performance beyond simple accuracy metrics.

Table 1. Classification report for RF Algorithm on Clinical Preoperative Data

	Precision	Recall	F1-score	Support
0	0.61	0.75	0.67	76
1	0.54	0.37	0.44	59
Accuracy			0.59	135
Macro avg	0.57	0.56	0.56	135
Weighted avg	0.58	0.59	0.57	135

Furthermore, delving into the feature importance analysis, a crucial aspect of understanding the model's predictive capabilities, Age emerges as the most influential variable in predicting MGMT promoter methylation status. The detailed breakdown of feature importance is visually represented in Fig. 3, providing a comprehensive overview of each variable's contribution to the RF model's predictions.

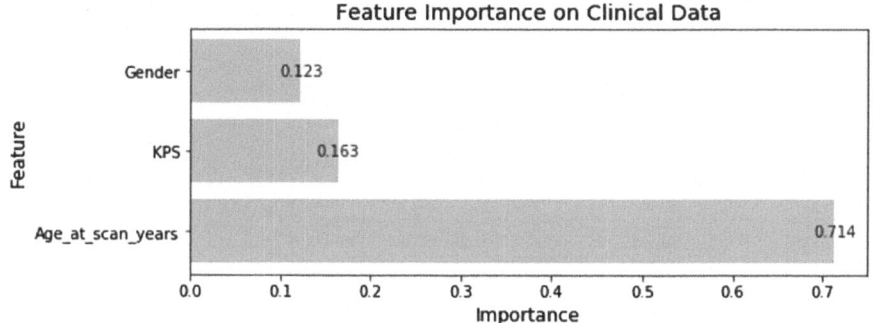

Fig. 3. Feature Importance Bar Plot on Clinical Preoperative Data.

In the context of acquisition data, with an overall accuracy of 59.3%, the RF model applied to the acquisition data demonstrates a reasonable ability to discriminate between the two classes. The confusion matrix in Fig. 4 indicates that 51 instances of non-methylation and 29 instances of methylation are correctly predicted, while 30 cases of methylation and 25 cases of non-methylation are misclassified. This results in an accuracy of 59.3%, with precision and recall values for both classes indicating a balanced predictive performance.

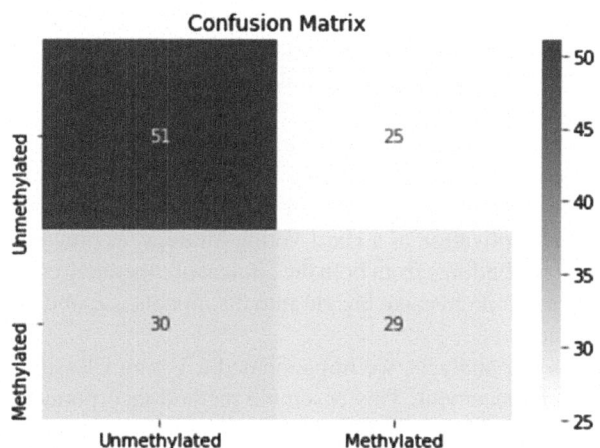

Fig. 4. Confusion Matrix for RF Algorithm on Acquisition Data.

Meanwhile, the classification report in Table 2 provides a detailed breakdown, showing that the model achieves precision, recall, and F1-score values of 0.63, 0.67, and 0.65 for the non-methylation class, and 0.54, 0.49, and 0.51 for the methylation class, respectively. These metrics offer insights into the model's ability to correctly identify instances of each class and highlight its performance in terms of precision (positive predictive value), recall (sensitivity), and the harmonic mean of precision and recall (F1-score).

Table 2. Classification report for RF Algorithm on Acquisition Data

	Precision	Recall	F1-score	Support
0	0.63	0.67	0.65	76
1	0.54	0.49	0.51	59
Accuracy			0.59	135
Macro avg	0.58	0.58	0.58	135
Weighted avg	0.59	0.59	0.59	135

It is noteworthy that in the process of predicting MGMT methylation status using the RF Classifier with Recursive Feature Elimination and Cross-Validation (RFECV), the algorithm identified T1 Imaging Frequency as the most predictive variable. The frequency of T1 imaging, a measure related to the characteristics of the MRI scans, emerges as a critical factor in predicting MGMT methylation status. This suggests that specific patterns or characteristics captured by T1 imaging frequency play a pivotal role in distinguishing between methylated and unmethylated MGMT states. This singular variable has demonstrated noteworthy significance in distinguishing MGMT methylation statuses. The feature importance analysis reveals that T1 Imaging Frequency carries substantial weight in determining the outcome of the model.

The identification of T1 Imaging Frequency as the most predictive variable opens avenues for further exploration. Future studies may delve into the specific imaging characteristics captured by this variable and their radiological relevance in the context of MGMT methylation.

In summary, the acquisition data model exhibits a comparable performance to the clinical preoperative data model, showcasing its potential utility in predicting MGMT promoter methylation status. The predictive capabilities of the models are further enhanced through the application of a Hard Voting Strategy, as previously stated. Hereafter, we will present the findings from both the clinical preoperative data and the acquisition data, providing a comprehensive insight into the models' capabilities and collective predictive performance.

As a final step in our analysis, we implemented a Voting Classifier using a resampling approach with replacement. This ensemble method incorporated Random Forest (RF), Extreme Gradient Boosting (XGB), and Support Vector Machine (SVM) models, showcasing robust performance with an average accuracy of 85.9%. The narrow standard deviation of accuracy at 0.8% indicates a consistent and reliable predictive capability. These results underscore the efficacy of our ensemble approach in achieving a high level of accuracy across diverse datasets and models.

The confusion matrix presented in Fig. 5 illustrates the detailed performance of the Voting Classifier. The model correctly predicted 72 instances of non-methylation and 44 instances of methylation. It demonstrated a remarkable ability to distinguish between the two classes, with only 4 instances of non-methylation and 15 instances of methylation being misclassified. These results highlight the enhanced predictive power achieved through the integration of diverse models and resampling techniques, emphasizing the robustness and reliability of the ensemble approach.

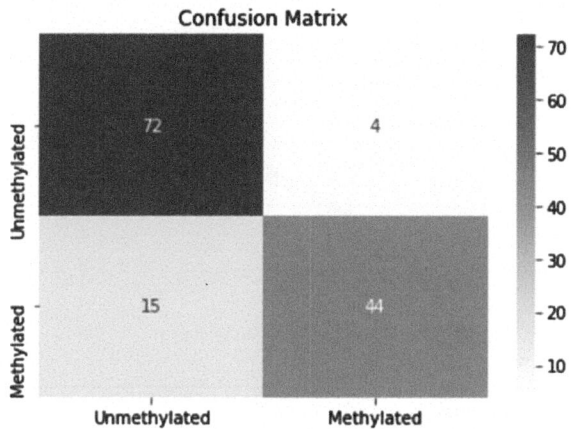

Fig. 5. Confusion Matrix for Hard Voting Classifier on Stacked Dataset.

5 Key Findings

In the initial model utilizing Random Forest on clinical data, the analysis of feature importance reports identified Age at scan years as the most influential variable in predicting the MGMT promoter methylation status. The model assigned the highest importance score to Age at scan years, highlighting its pivotal role in distinguishing between MGMT methylated and non-methylated cases. Alongside Gender and KPS (Karnofsky Performance Status), Age at scan years emerged as a crucial factor contributing to the predictive power of the model.

To assess the statistical significance of the Age difference between the MGMT methylated and non-methylated groups, a non-parametric test, specifically the Mann-Whitney U test, was employed. The U-Statistic obtained was 11902.500, and the corresponding p-value (Mann-Whitney U) was 0.022. This result indicates that the difference in Age between MGMT methylated and non-methylated groups is statistically significant.

It is noteworthy that the choice of the Mann-Whitney U test was based on the non-normality of the Age distribution, as confirmed by the Shapiro-Wilk test (p-value: 2.5783385353861377e-05), is reinforced by the Q-Q (Quantile-Quantile) plot in Fig. 6, providing visual evidence of the deviation from normality. This statistical test is particularly suitable when dealing with non-normally distributed data, and it is applicable to both continuous and categorical variables. Given that the MGMT variable is categorical, representing two distinct groups, the Mann-Whitney U test aligns well with the nature of the data. This non-parametric test provides a robust approach for assessing differences in the distribution of Age between the two MGMT groups, offering reliable results even when normality assumptions are not met.

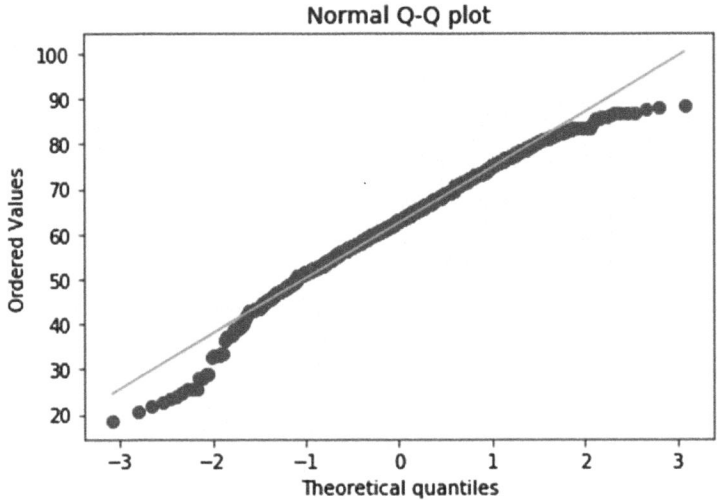

Fig. 6. Q-Q Plot of Age Distribution.

The combination of these statistical analyses and the alignment with existing literature [25–27] underscores the importance of Age as a relevant factor in the context of MGMT methylation status. This information enhances the overall understanding of the predictive capabilities of the model and contributes valuable insights to the field of glioblastoma research.

In the second model applied to the acquisition data, the Feature Importance analysis revealed T1 Imaging Frequency as the most predictive variable for determining the MGMT promoter methylation status. This finding aligns with previous studies that demonstrated the association between individual MR features and MGMT methylation [28, 29]. It is evident that better algorithms are needed to enhance the predictive accuracy of MGMT methylation status, reflecting the ongoing challenges in glioblastoma research. The study's significance extends to the application of novel techniques for anatomical MR image analysis, allowing for a more nuanced evaluation of features that might escape human observation.

In summary, the combination of clinical features like Gender, KPS, and Age at scan years, alongside T1 Imaging Frequency from acquisition data, collectively play pivotal roles in predicting MGMT promoter methylation status in glioblastoma. This underscores the significance of incorporating diverse datasets to enhance predictive accuracy. Further advancements in algorithm development are crucial to refine MGMT status prediction and maximize the potential of non-invasive diagnostic approaches. The study's exploration of novel anatomical MR image analysis techniques, alongside the integration of clinical and acquisition data, marks a significant shift towards reducing the reliance on invasive procedures in glioblastoma management.

6 Conclusion

In conclusion, our investigation into predicting MGMT promoter methylation status not only encompassed distinct datasets, clinical preoperative and acquisition, but also emphasized the potency of a Hard Voting Strategy. This ensemble approach, combining the strengths of individual models, was implemented with the overarching goal of achieving non-invasive and more accurate predictions. The findings underscore the potential of integrated and robust predictive models, laying the foundation for advancements in diagnostic assessments and therapeutic strategies in GBM research.

This research establishes a framework for advancing diagnostic assessments and therapeutic strategies by minimizing reliance on invasive procedures. Through the integration of diverse datasets and collaborative efforts with multidisciplinary teams, it offers a comprehensive approach to understanding and addressing GBM, particularly in the context of MGMT prediction. By reducing the need for invasive procedures, this approach reshapes the landscape of diagnosis and treatment, offering more patient-friendly alternatives and enhancing overall care for GBM patients.

References

1. Brown, N., et al.: Survival outcomes and prognostic factors in glioblastoma. Cancers **14**(13) (2022). https://doi.org/10.3390/cancers14133161
2. Chai, R., et al.: Predictive value of MGMT promoter methylation on the survival of TMZ treated IDH-mutant glioblastoma. Cancer Bio. Med. **18**(1), 272–282 (2021). https://doi.org/10.20892/J.ISSN.2095-3941.2020.0179
3. von Känel, T., Huber, A.: DNA methylation analysis. Swiss Med. Weekly **143**(2122) (2013). https://doi.org/10.4414/SMW.2013.13799
4. Mansouri, A., et al.: MGMT promoter methylation status testing to guide therapy for glioblastoma: refining the approach based on emerging evidence and current challenges. Neuro Oncol. **21**(2), 167–178 (2019). https://doi.org/10.1093/neuonc/noy132
5. Saeed, N., et al.: Is it possible to predict MGMT promoter methylation from Brain Tumor MRI scans using deep learning models? PMLR Proc. Mach. Learn. Res. **172**, 1005–1018 (2022)
6. Saeed, N., et al.: MGMT promoter methylation status prediction using MRI scans? an extensive experimental evaluation of deep learning models. Med. Image Anal. **90** (2023). https://doi.org/10.1016/j.media.2023.102989
7. Sidey-Gibbons, J., et al.: Machine learning in medicine: a practical introduction. BMC Med. Res. Methodol. **19** (2019). https://doi.org/10.1186/S12874-019-0681-4
8. Javaid, M., et al.: Significance of machine learning in healthcare: features, pillars and applications. Int. J. Intell. Netw. **3**, 58–73 (2022). https://doi.org/10.1016/j.ijin.2022.05.002
9. Althnian, A., et al.: Impact of dataset size on classification performance: an empirical evaluation in the medical domain. Appl. Sci. **11**(2) (2021). https://doi.org/10.3390/APP11020796
10. Lugg-Widger, F., et al.: Challenges in accessing routinely collected data from multiple providers in the UK for primary studies: managing the morass. Int. J. Popul. Data Sci. **3**(3) (2018). https://ijpds.org/article/view/432
11. Furstenau, L., et al.: Big data in healthcare: conceptual network structure, key challenges and opportunities. Dig. Commun. Netw. **9**(4), 856–868 (2023). https://doi.org/10.1016/j.dcan.2023.03.005

12. Bakas, S., et al.: The University of Pennsylvania glioblastoma (UPenn-GBM) cohort: advanced MRI, clinical, genomics, and radiomics. Sci. Data **9** (2022). https://doi.org/10.1038/s41597-022-01560-7
13. Jiang, H., et al.: Foundation of preoperative prognosis estimation model for glioblastoma multiforme. National Med. J. China **97**(31) (2017). https://doi.org/10.3760/cma.j.issn.0376-2491.2017.31.013
14. Pierscianek, D., et al.: Preoperative survival prediction in patients with glioblastoma by routine inflammatory laboratory parameters. Anticancer Res. **40**(2), 1161–1166 (2020). https://doi.org/10.21873/anticanres.14058
15. Mirpuri, P., et al.: The association of preoperative frailty and neighborhood-level disadvantage with outcome in patients with newly diagnosed glioblastoma. World Neurosurg. **166**, e949–e957 (2022). https://doi.org/10.1016/j.wneu.2022.07.138
16. Péus, D., et al.: Appraisal of the Karnofsky performance Status and proposal of a simple algorithmic system for its evaluation. BMC Med. Inform. Decis. Making **13** (2013). https://doi.org/10.1186/1472-6947-13-72
17. Schag, C., et al.: Karnofsky performance status revisited: reliability, validity, and guidelines. J. Clin. Oncol. **2**(3) (1984). https://doi.org/10.1200/JCO.1984.2.3.187
18. Ellis, R.J., Sander, R.M., Limon, A.: Twelve key challenges in medical machine learning and solutions. Intell.-Based Med. **6**, 100068 (2022). https://doi.org/10.1016/j.ibmed.2022.100068, ISSN 2666-5212
19. Le, N., et al.: XGBoost improves classification of MGMT promoter methylation status in IDH1 wildtype glioblastoma. J. Personalized Med. **10**(3) (2020). https://doi.org/10.3390/JPM10030128
20. Saxena, S., et al.: Role of artificial intelligence in Radiogenomics for cancers in the era of precision medicine. Cancers **14**(12) (2022). https://doi.org/10.3390/cancers14122860
21. Joshi, R., et al.: Ensemble based machine learning approach for prediction of glioma and multi-grade classification. Comput. Biol. Med. **137** (2021). https://doi.org/10.1016/j.compbiomed.2021.104829
22. Breiman, L.: Bagging predictors. Mach. Learn. **24**, 123–140 (1996). https://doi.org/10.1023/A:1018054314350
23. Leon, F., et al.: Evaluating the effect of voting methods on ensemble-based classification. In: IEEE International Conference on INnovations in Intelligent SysTems and Applications (INISTA), Gdynia, Poland, pp. 1–6 (2017). https://doi.org/10.1109/inista.2017.8001122
24. Pedregosa, F., et al.: Scikit-learn: machine learning in Python. J. Mach. Learn. Res. **12**, 2825–2830 (2011)
25. Horbinski, C., et al.: MGMT promoter methylation is associated with patient age and 1p/19q status in IDH-mutant gliomas. Neuro-Oncol. **23** (5) (2021). https://doi.org/10.1093/neuonc/noab039
26. Gerstner, E., et al.: MGMT methylation status may predict survival in elderly patients with newly diagnosed glioblastoma (GBM). J. Clin. Oncol. **27**(5_suppl) (2009). https://doi.org/10.1200/jco.2009.27.15_suppl.e13023
27. Yin, A., et al.: The predictive but not prognostic value of MGMT promoter methylation status in elderly glioblastoma patients: a meta-analysis. PLoS ONE **9**(1), e85102 (2014). https://doi.org/10.1371/journal.pone.0085102
28. Drabycz, S., et al.: An analysis of image texture, tumor location, and MGMT promoter methylation in glioblastoma using magnetic resonance imaging. Neuroimage **49**(2), 1398–1405 (2010). https://doi.org/10.1016/j.neuroimage.2009.09.049
29. Qureshi, S., et al.: Radiogenomic classification for MGMT promoter methylation status using multi-omics fused feature space for least invasive diagnosis through mpMRI scans. Sci. Reports **13** (2023). https://doi.org/10.1038/s41598-023-30309-4

Author Index

A
AbouHassan, Iman 64
Akmal, Muhammad Uzair 128
Anastasiadis, Dimitrios 211
Asif, Saara 128

B
Bădică, Costin 199
Brito, José 80

C
Chen, Liming 12

D
Dhelim, Sahraoui 12
Dimitrov, Dimitar 1

G
Georgieva, Petia 80
Grancharova, Alexandra 155
Grossmann, Daniel 128

H
Hamady, Mohamad 93

I
Ilie, Sorin 199

J
Janjani, Pranav 168

K
Kamar, Issa 93
Karakehayova, Alexandrina 40
Kasabov, Nikola 64
Kazi, Faruk 168
Kechadi, M-Tahar 12
Knollmeyer, Simon 128

Kocijan, Juš 155
Koprinkova-Hristova, Petia 54
Koval, Leonid 128
Koychev, Ivan 1

L
Li, Yunan 12
Lyubchev, Dimitar 104

M
Marchev Jr., Angel 104
Marfurt, Andreas 184
Markov, Borislav 54
Markov, Konstantin 143
Mathias, Selvine G. 128
Mihov, Stoyan 27
Miteva, Mariya 223
Momcheva, Galina 93
Murarețu, Ionuț 199

N
Nabeel, Aayad 93
Nakov, Preslav 1
Nisheva-Pavlova, Maria 40, 223

O
Oğul, Hasan 184

P
Palan, Mayank 168
Penchev, Nikolay 104
Popov, George 64

R
Ragheb, Mostafa 93
Refanidis, Ioannis 211
Rieger, Ines 122

S
Santana, Filipa 80
Shegokar, Ninad 168
Shirude, Sarvesh 168

T
Tanner, Thomas 184
Trifonov, Roumen 64

V
Vankov, Georgi 1
Vultureanu-Albişi, Alexandra 199

X
Xie, Junhong 155

Z
Zou, Xinhao 143

The manufacturer's authorised representative in the EU is Springer Nature Customer Service Centre GmbH, Europaplatz 3, 69115 Heidelberg, Germany. If you have any concerns regarding our products, please contact ProductSafety@springernature.com

Printed and bound by CPI Group (UK) Ltd, Croydon, CR0 4YY

26/03/2026

02078952-0005